JUMP Math 2.1
Book 2 Part 1 of 2

Contents

jump math™
MULTIPLYING POTENTIAL.

JUMP Math
One Yonge Street, Suite 1014
Toronto, Ontario M5E 1E5
Canada
www.jumpmath.org

Writers: Dr. Heather Betel, Julie Lorinc, Dr. John Mighton
Consultants: Dr. Anna Klebanov, Dr. Sindi Sabourin
Editors: Megan Burns, Liane Tsui, Natalie Francis, Annie Chern, Julia Cochrane, Janice Dyer, Dawn Hunter,
        Neomi Majmudar, Una Malcolm, Rita Vanden Heuvel
Layout and Illustrations: Linh Lam, Gabriella Kerr, Pam Lostracco
Cover Design: Blakeley Words+Pictures
Cover Photograph: © iStockphoto.com/Michael Valdez

ISBN 978-1-927457-37-5

Second printing July 2017

Printed and bound in Canada

Welcome to JUMP Math

Entering the world of JUMP Math means believing that every child has the capacity to be fully numerate and to love math. Founder and mathematician John Mighton has used this premise to develop his innovative teaching method. The resulting resources isolate and describe concepts so clearly and incrementally that everyone can understand them.

JUMP Math is comprised of teacher's guides (which are the heart of our program), interactive whiteboard lessons, student assessment & practice books, evaluation materials, outreach programs, and teacher training. The Common Core Editions of our resources have been carefully designed to cover the Common Core State Standards. All of this is presented on the JUMP Math website: **www.jumpmath.org**.

Teacher's guides are available on the website for free use. Read the introduction to the teacher's guides before you begin using these resources. This will ensure that you understand both the philosophy and the methodology of JUMP Math. The assessment & practice books are designed for use by students, with adult guidance. Each student will have unique needs and it is important to provide the student with the appropriate support and encouragement as he or she works through the material.

Allow students to discover the concepts by themselves as much as possible. Mathematical discoveries can be made in small, incremental steps. The discovery of a new step is like untangling the parts of a puzzle. It is exciting and rewarding.

Students will need to answer the questions marked with a 📓 in a notebook. Grid paper notebooks should always be on hand for answering extra questions or when additional room for calculation is needed.

Contents

Unit 4: Operations and Algebraic Thinking: Unknowns in Subtraction

Unit 5: Number and Operations in Base Ten: Addition Using Place Value

Unit 6: Number and Operations in Base Ten: Subtraction Using Place Value

Unit 7: Measurement and Data: Measuring Length in Metric Units

Unit 8: Measurement and Data: Measuring and Operations

PART 2
Unit 1: Operations and Algebraic Thinking: Compare Problems

Unit 2: Number and Operations in Base Ten: Three-Digit Numbers

Unit 3: Operations and Algebraic Thinking: Two-Step Word Problems

Unit 4: Number and Operations in Base Ten: Strategies for Large Numbers

Unit 5: Measurement and Data: Measuring in US Customary Units

Unit 6: Measurement and Data: Time

Unit 7: Measurement and Data: Money

Unit 8: Geometry: Shapes

Unit 9: Measurement and Data: Graphs

OA2-1 The Next Number

☐ Write the next number.

1. 4, __5__

2. 2, _____

3. 8, _____

4. 5, _____

5. 3, _____

6. 7, _____

7. 6, _____

8. 1, _____

☐ Write the next number.

9. 3, __4__
23, __24__

10. 7, _____
47, _____

11. 5, _____
35, _____

12. 2, _____
72, _____

13. 8, _____
18, _____

14. 6, _____
86, _____

15. 4, _____
54, _____

16. 1, _____
91, _____

☐ Write the next two numbers.

17. 2, __3__, __4__
52, __53__, __54__

18. 1, _____, _____
21, _____, _____

19. 7, _____, _____
47, _____, _____

20. 3, _____, _____
93, _____, _____

21. 6, _____, _____
76, _____, _____

22. 5, _____, _____
85, _____, _____

○ Write the next three numbers.

23.

3, __4__ , __5__ , __6__   23, __24__ , __25__ , __26__   63, __64__ , __65__ , __66__

24.

16, ____ , ____ , ____   36, ____ , ____ , ____   76, ____ , ____ , ____

25.

24, ____ , ____ , ____   54, ____ , ____ , ____   84, ____ , ____ , ____

26.

1, ____ , ____ , ____   41, ____ , ____ , ____   91, ____ , ____ , ____

27.

15, ____ , ____ , ____   65, ____ , ____ , ____   75, ____ , ____ , ____

○ Write the next four numbers.

28.

25, ____ , ____ , ____ , ____   45, ____ , ____ , ____ , ____

29.

13, ____ , ____ , ____ , ____   73, ____ , ____ , ____ , ____

30.

41, ____ , ____ , ____ , ____   91, ____ , ____ , ____ , ____

31.

2, ____ , ____ , ____ , ____   82, ____ , ____ , ____ , ____

OA2-2 Addition

☐ Add.

1.
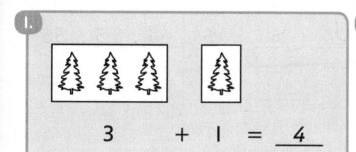

3 + 1 = __4__

2.
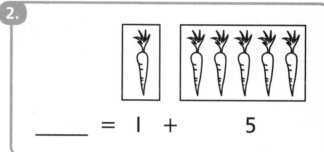

____ = 1 + 5

3.
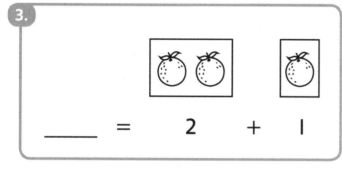

____ = 2 + 1

4.
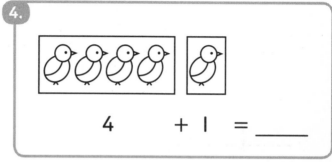

4 + 1 = ____

5.
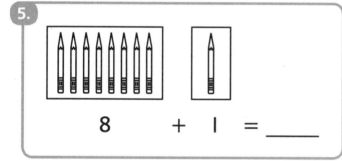

8 + 1 = ____

6.
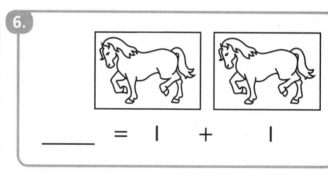

____ = 1 + 1

7.
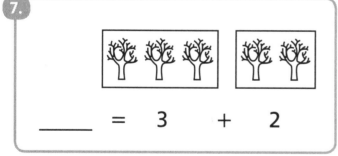

____ = 3 + 2

8.
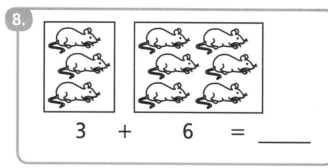

3 + 6 = ____

9.
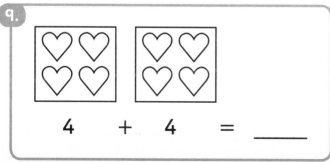

4 + 4 = ____

10.
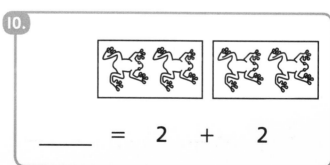

____ = 2 + 2

⬜ Add.

11.
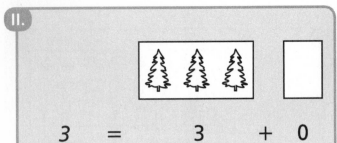

___3___ = 3 + 0

12.
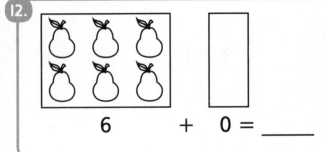

6 + 0 = ___

13.
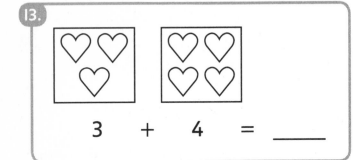

3 + 4 = ___

14.
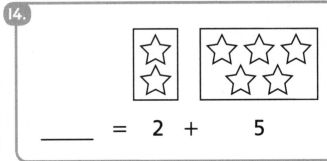

___ = 2 + 5

15.
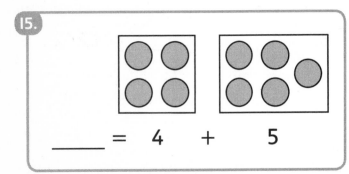

___ = 4 + 5

16.
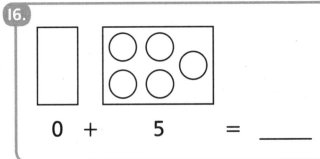

0 + 5 = ___

17.
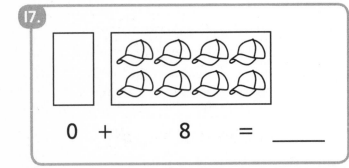

0 + 8 = ___

18.
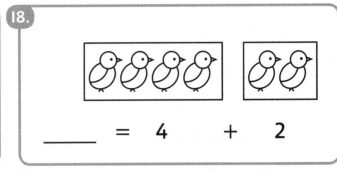

___ = 4 + 2

19.
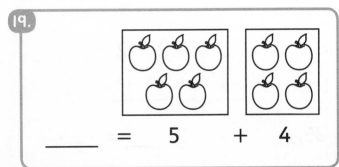

___ = 5 + 4

20.
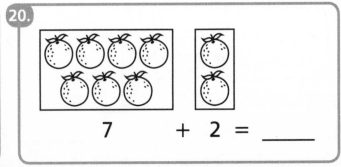

7 + 2 = ___

Operations and Algebraic Thinking 2-2

OA2-3 Adding by Counting On

☐ Add 1 by counting on.

1.

$3 + 1 =$ _____

3   4

so $3 + 1 =$ __4__

$13 + 1 =$ _____

13   14

so $13 + 1 =$ __14__

2.

$7 + 1 =$ _____ $47 + 1 =$ _____ $87 + 1 =$ _____

3.

_____ $= 2 + 1$ _____ $= 32 + 1$ _____ $= 62 + 1$

☐ Add 2 by counting on.

4.

$6 + 2 =$ _____

6   7   8

so $6 + 2 =$ __8__

$56 + 2 =$ _____

56   57   58

so $56 + 2 =$ __58__

5.

$3 + 2 =$ _____ $53 + 2 =$ _____ $63 + 2 =$ _____

6.

_____ $= 15 + 2$ _____ $= 35 + 2$ _____ $= 75 + 2$

◯ Add 3 by counting on.

7.

$2 + 3 =$ _____ $\qquad\qquad$ $12 + 3 =$ _____

2 $\qquad$ 3 $\qquad$ 4 $\qquad$ 5 $\qquad\qquad$ 12 $\qquad$ 13 $\qquad$ 14 $\qquad$ 15

so $2 + 3 =$ __5__ $\qquad\qquad$ so $12 + 3 =$ __15__

8.

$4 + 3 =$ _____ $\qquad$ $54 + 3 =$ _____ $\qquad$ $74 + 3 =$ _____

9.

_____ $= 21 + 3$ $\qquad$ _____ $= 41 + 3$ $\qquad$ _____ $= 81 + 3$

10.

$5 + 3 =$ _____ $\qquad$ $55 + 3 =$ _____ $\qquad$ $65 + 3 =$ _____

11.

_____ $= 6 + 3$ $\qquad$ _____ $= 46 + 3$ $\qquad$ _____ $= 96 + 3$

12. BONUS

$110 + 3 =$ __113__ $\qquad$ $140 + 3 =$ _____ $\qquad$ $170 + 3 =$ _____

13. BONUS

_____ $= 230 + 3$ $\qquad$ _____ $= 250 + 3$ $\qquad$ _____ $= 280 + 3$

OA2-4 Even and Odd Numbers

☐ Count the stars.
☐ Pair up as many stars as you can.
☐ Write **even** if you can pair all the stars.
  Write **odd** if you cannot.

1.

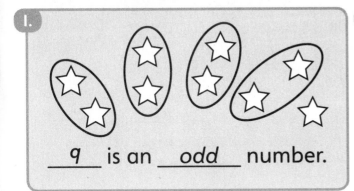

___9___ is an ___odd___ number.

2.

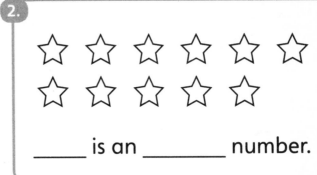

_____ is an _____ number.

3.

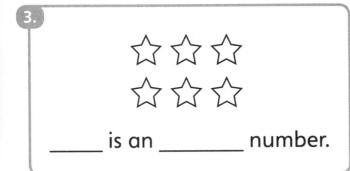

_____ is an _____ number.

4.

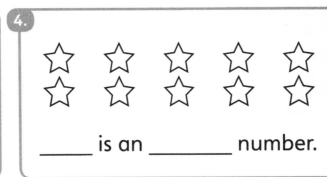

_____ is an _____ number.

5.

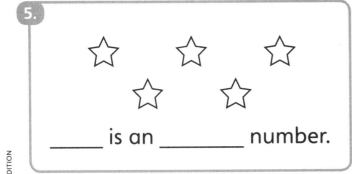

_____ is an _____ number.

6.

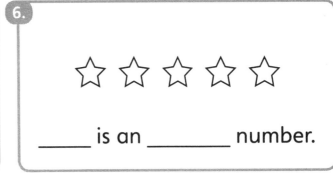

_____ is an _____ number.

7.

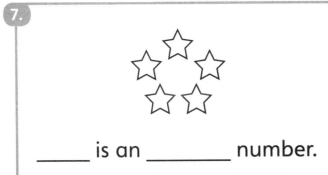

_____ is an _____ number.

8.

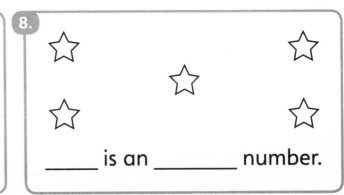

_____ is an _____ number.

◻ Draw a line to divide the dots into 2 equal groups if you can.

◻ Write **even** if you can divide the dots equally.
Write **odd** if you cannot.

9.

8 is ___even___.

10.

7 is ___odd___.

11.

6 is _____.

12.

3 is _____.

13.

___ is _____.

14.

___ is _____.

15.

___ is _____.

16.

___ is _____.

17.

___ is _____.

Operations and Algebraic Thinking 2-4

OA2-5 Patterns with Even and Odd Numbers

☐ Pair the faces up.

☐ Write **even** if you can pair all the objects.
   Write **odd** if you cannot.

1.
☺

1 is _____.

2.
☺ ☺

2 is _____.

3.
☺ ☺ ☺

3 is _____.

4.
☺ ☺
☺ ☺

4 is _____.

5.
☺ ☺ ☺
☺ ☺

5 is _____.

6.
☺ ☺ ☺
☺ ☺ ☺

6 is _____.

7.
☺ ☺ ☺ ☺
☺ ☺ ☺

7 is _____.

8.
☺ ☺ ☺ ☺
☺ ☺ ☺ ☺

8 is _____.

9.
☺ ☺ ☺ ☺ ☺
☺ ☺ ☺ ☺

9 is _____.

☐ Write **O** for odd and **E** for even.

☐ Extend both patterns.

10.

1	2	3	4	5	6	7	8	9	10
_O_	_E_	___	___	___	___	___	___	___	___

11	12	13	14	15	16				
						___	___	___	___

___ ___ ___ ___ ___ ___ ___ ___ ___ ___

☐ Shade the even numbers.
☐ Circle the odd numbers.

11.

1	2	3	4	5	6	7	8	9	10
11	12	13	14	15	16	17	18	19	20
21	22	23	24	25	26	27	28	29	30

☐ Write the even numbers.

12.

___ ___ ___ ___ ___

___ ___ ___ ___ ___

___ ___ ___ ___ ___

☐ Write the odd numbers.

13.

___ ___ ___ ___ ___

___ ___ ___ ___ ___

___ ___ ___ ___ ___

14.

Even numbers end with ____, ____, ____, ____, or ____.

Odd numbers end with ____, ____, ____, ____, or ____.

☐ Circle the even numbers.
☐ Underline the odd numbers.

15.

<u>5</u>   ②   8   1   6   4   3   7   10   9

Operations and Algebraic Thinking 2-5

OA2-6 The Next or Previous Even or Odd Number

☐ Write the next even number.

1. 6, _____

2. 8, _____

3. 4, _____

4. 2, _____

5. 16, _____

6. 24, _____

7. 20, _____

8. 14, _____

☐ Write the next odd number.

9. 7, _____

10. 1, _____

11. 5, _____

12. 3, _____

13. 11, _____

14. 23, _____

15. 27, _____

16. 15, _____

☐ Write the even number before.

17. _____, 6

18. _____, 4

19. _____, 8

20. _____, 10

21. _____, 22

22. _____, 12

23. _____, 26

24. _____, 16

☐ Write the odd number before.

25. _____, 9

26. _____, 3

27. _____, 5

28. _____, 7

29. _____, 13

30. _____, 29

31. _____, 17

32. _____, 25

☐ Write the next even number.
☐ Write an addition sentence to show adding 2.

33.
6, _8_   _6_ + _2_ = _8_

34.
4, ____   ____ + ____ = ____

35.
10, ____   ____ + ____ = ____

36.
12, ____   ____ + ____ = ____

☐ Write the next even number.

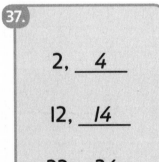

37.
2, _4_

12, _14_

22, _24_

32, _34_

42, _44_

52, _54_

62, _64_

72, _74_

82, _84_

92, _94_

38.
6, ____

16, ____

26, ____

36, ____

46, ____

56, ____

66, ____

76, ____

86, ____

96, ____

39.
4, ____

14, ____

24, ____

34, ____

44, ____

54, ____

64, ____

74, ____

84, ____

94, ____

40.
8, _10_

18, _20_

28, ____

38, ____

48, ____

58, ____

68, ____

78, ____

88, ____

98, ____

○ Write the next odd number.

○ Write an addition sentence to show adding 2.

41.

5, _7_   _5_ + _2_ = _7_

42.

3, ____   ____ + ____ = ____

43.

11, ____   ____ + ____ = ____

44.

17, ____   ____ + ____ = ____

○ Write the next odd number.

45.

3, _5_

13, _15_

23, _25_

33, _35_

46.

7, ____

17, ____

27, ____

37, ____

47.

9, _11_

19, ____

29, ____

39, ____

48.

5, ____

15, ____

25, ____

35, ____

○ Do you add 1 or 2 to get the answer?

49.

5 + _1_ = 6

50.

2 + _2_ = 4

51.

7 + ____ = 9

52.

16 + ____ = 18

53.

15 + ____ = 16

54.

8 + ____ = 9

55.

10 + ____ = 11

56.

11 + ____ = 13

57.

18 + ____ = 20

Operations and Algebraic Thinking 2-6

OA2-7 Order in Adding

How many dots are on each side of the domino?
☐ Add the dots to find the total.

1.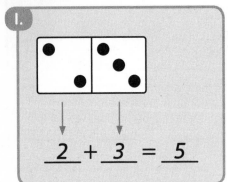

2 + 3 = 5

2.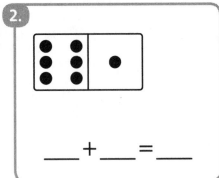

___ + ___ = ___

3.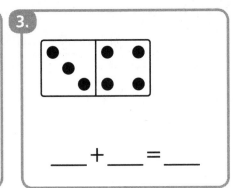

___ + ___ = ___

Ben turned the domino around.
Did the number of dots change?
☐ Circle **Yes** or **No**.

4.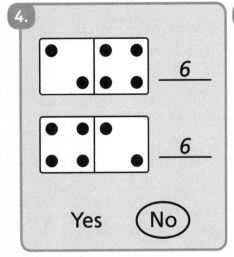

6

6

Yes   (No)

5.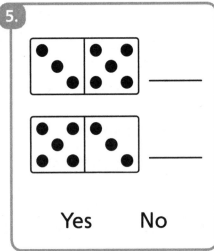

___

___

Yes   No

6.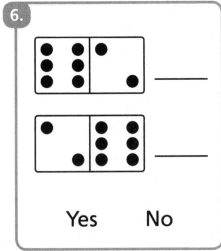

___

___

Yes   No

7.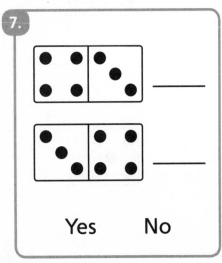

___

___

Yes   No

8.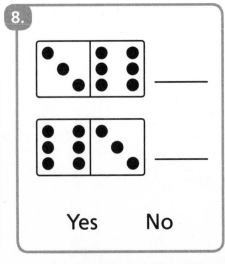

___

___

Yes   No

9.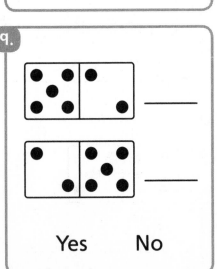

___

___

Yes   No

Rosa turned the domino around.
☐ Write two addition sentences.

10.

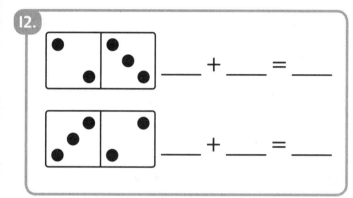

$\underline{2} + \underline{1} = \underline{3}$

$\underline{1} + \underline{2} = \underline{3}$

11.

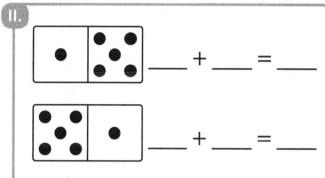

$\underline{} + \underline{} = \underline{}$

$\underline{} + \underline{} = \underline{}$

12.

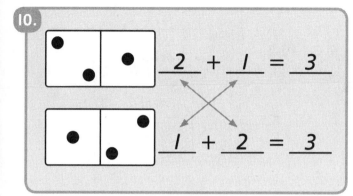

$\underline{} + \underline{} = \underline{}$

$\underline{} + \underline{} = \underline{}$

13.

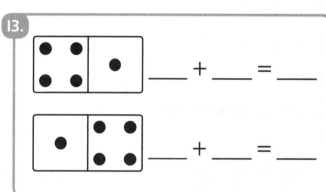

$\underline{} + \underline{} = \underline{}$

$\underline{} + \underline{} = \underline{}$

Ali turned the domino around.
☐ Add the dots on the first domino.
☐ How many dots are on the other domino?

14.

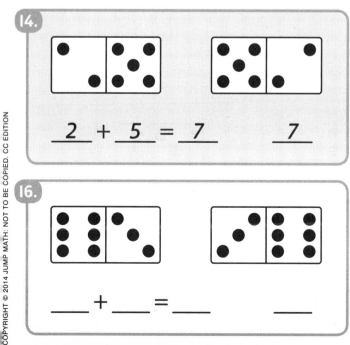

$\underline{2} + \underline{5} = \underline{7} \qquad \underline{7}$

15.

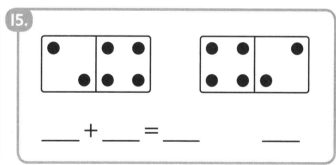

$\underline{} + \underline{} = \underline{} \qquad \underline{}$

16.

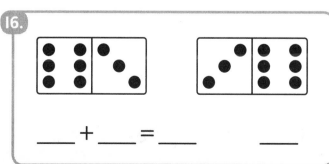

$\underline{} + \underline{} = \underline{} \qquad \underline{}$

17.

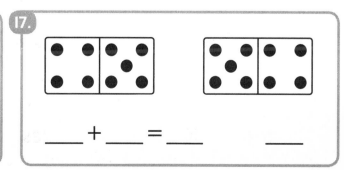

$\underline{} + \underline{} = \underline{} \qquad \underline{}$

⬜ Add.

⬜ Change the order of the numbers.

⬜ Add again.

⬜ Did the answer change? Circle **Yes** or **No**.

18.

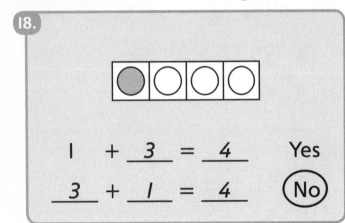

$$1 + \underline{\ 3\ } = \underline{\ 4\ } \qquad \text{Yes}$$
$$\underline{\ 3\ } + \underline{\ 1\ } = \underline{\ 4\ } \qquad \boxed{\text{No}}$$

19.

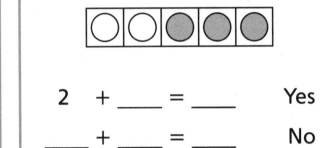

$$2 + \underline{\quad} = \underline{\quad} \qquad \text{Yes}$$
$$\underline{\quad} + \underline{\quad} = \underline{\quad} \qquad \text{No}$$

20.

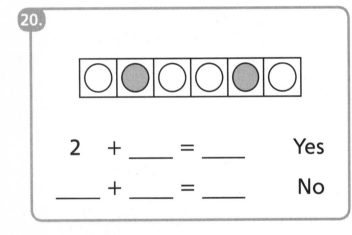

$$2 + \underline{\quad} = \underline{\quad} \qquad \text{Yes}$$
$$\underline{\quad} + \underline{\quad} = \underline{\quad} \qquad \text{No}$$

21.

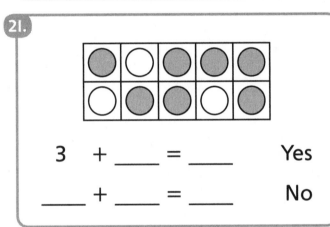

$$3 + \underline{\quad} = \underline{\quad} \qquad \text{Yes}$$
$$\underline{\quad} + \underline{\quad} = \underline{\quad} \qquad \text{No}$$

⬜ Change the order of the numbers.

⬜ Are the totals the same?

22.

$$5 + 1 = \underline{\ 6\ } \qquad \boxed{\text{Yes}}$$
$$\underline{\ 1\ } + \underline{\ 5\ } = \underline{\ 6\ } \qquad \text{No}$$

23.

$$5 + 2 = \underline{\quad} \qquad \text{Yes}$$
$$\underline{\quad} + \underline{\quad} = \underline{\quad} \qquad \text{No}$$

24.

$$7 + 1 = \underline{\quad} \qquad \text{Yes}$$
$$\underline{\quad} + \underline{\quad} = \underline{\quad} \qquad \text{No}$$

25.

$$4 + 2 = \underline{\quad} \qquad \text{Yes}$$
$$\underline{\quad} + \underline{\quad} = \underline{\quad} \qquad \text{No}$$

Operations and Algebraic Thinking 2-7

OA2-8 First Word Problems

☐ Add using the pictures.

1.

4 bunnies

2 more bunnies

4 + 2 = ____

2.

3 bunnies

1 more bunny

3 + 1 = ____

3.

1 bunny

5 more bunnies

1 + 5 = ____

4.

2 bunnies

6 more bunnies

2 + 6 = ____

Add using the pictures.

5.

3 big frogs   and   2 small frogs

3 + 2 = _____ frogs altogether

6.

I big frog   and   4 small frogs

I + 4 = _____ frogs altogether

7.

5 big frogs   and   3 small frogs

5 + 3 = _____ frogs altogether

8.

2 small frogs   and   7 big frogs

2 + 7 = _____ frogs altogether

☐ Draw circles to show the numbers.
☐ Add.

9.

There are 4 cats.   Then 2 more cats come.

___   ___   ___   ___   ___   ___

4 + 2 = _____ cats altogether

10.

4 large dogs   and   3 small dogs

___ ___ ___ ___ ___ ___ ___

4 + 3 = _____ dogs altogether

11.

There are 3 glasses of milk. Nina brings 2 more glasses of milk.

___ ___ ___   ___ ___

3 + 2 = _____ glasses altogether

12.

6 small birds   and   2 large birds

6 + 2 = _____ birds altogether

OA2-9 Equal (=) and Not Equal (≠)

Do the tables have the same number of balls?

☐ If they do, circle =.

☐ If they do not, circle ≠.

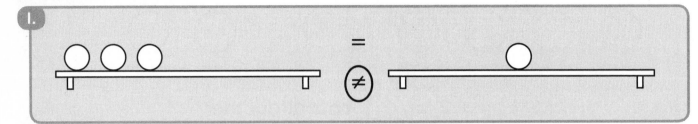

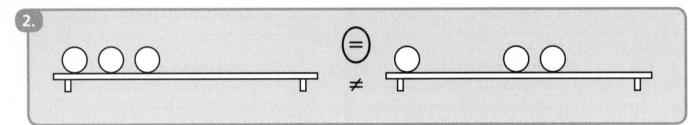

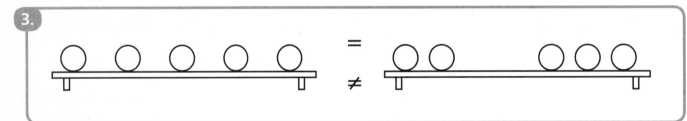

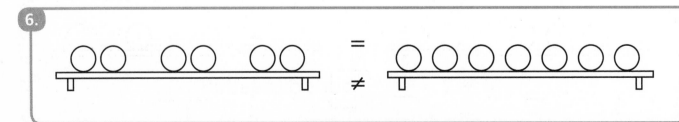

Operations and Algebraic Thinking 2-9

⬚ Write the number of balls.

⬚ Write = or ≠ in the box.

7.

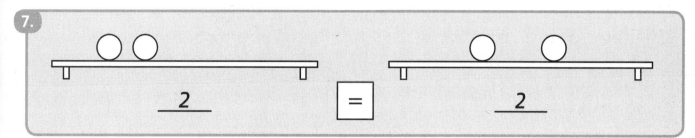

    _2_   =   _2_

8.

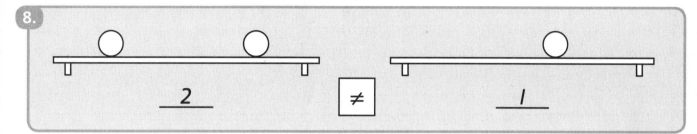

    _2_   ≠   _1_

9.

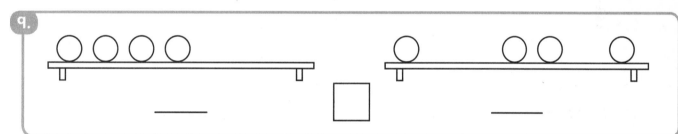

10.

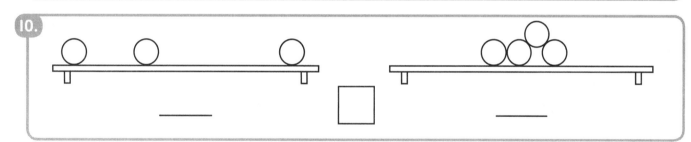

11.

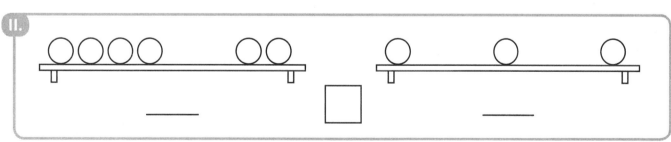

12.

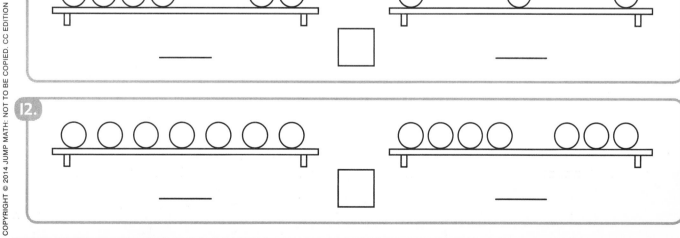

☐ Write the number of balls.

☐ Write = or ≠ in the box.

13.

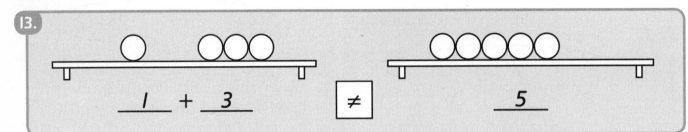

$\underline{\ \ 1\ \ } + \underline{\ \ 3\ \ }$　$\neq$　$\underline{\ \ 5\ \ }$

14.

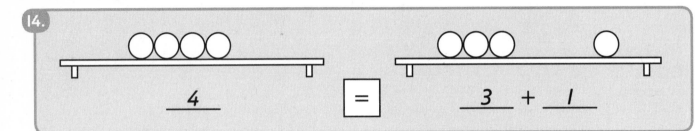

$\underline{\ \ 4\ \ }$　$=$　$\underline{\ \ 3\ \ } + \underline{\ \ 1\ \ }$

15.

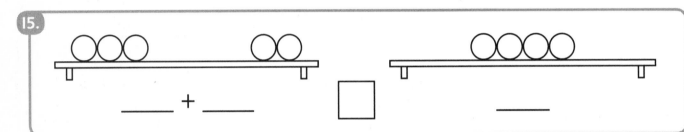

$\underline{\ \ \ \ } + \underline{\ \ \ \ }$　☐　$\underline{\ \ \ \ }$

16.

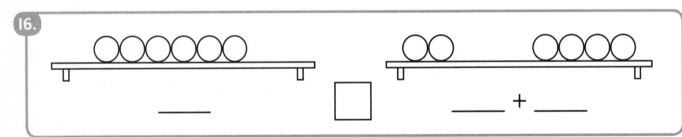

$\underline{\ \ \ \ }$　☐　$\underline{\ \ \ \ } + \underline{\ \ \ \ }$

17.

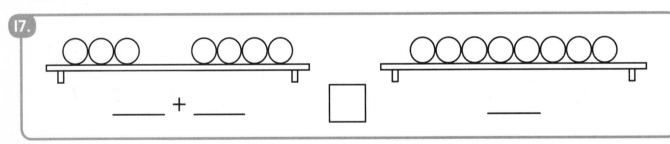

$\underline{\ \ \ \ } + \underline{\ \ \ \ }$　☐　$\underline{\ \ \ \ }$

18.

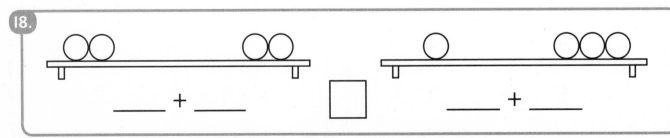

$\underline{\ \ \ \ } + \underline{\ \ \ \ }$　☐　$\underline{\ \ \ \ } + \underline{\ \ \ \ }$

Circle the correct addition sentence.

19.

$6 \quad = \quad 2 + 1$

$\boxed{6 \quad \neq \quad 2 + 1}$ ⬭

20.

⬭ $\boxed{4 \quad = \quad 2 + 2}$

$4 \quad \neq \quad 2 + 2$

21.

$2 + 1 \quad = \quad 3$

$2 + 1 \quad \neq \quad 3$

22.

$2 + 2 \quad = \quad 5$

$2 + 2 \quad \neq \quad 5$

23.

$6 \quad = \quad 5 + 1$

$6 \quad \neq \quad 5 + 1$

24.

$6 \quad = \quad 4 + 3$

$6 \quad \neq \quad 4 + 3$

25.

$8 + 2 \quad = \quad 9$

$8 + 2 \quad \neq \quad 9$

26.

$3 + 5 \quad = \quad 8$

$3 + 5 \quad \neq \quad 8$

27.

$10 \quad = \quad 4 + 6$

$10 \quad \neq \quad 4 + 6$

28.

$7 \quad = \quad 2 + 7$

$7 \quad \neq \quad 2 + 7$

29.

$4 + 5 \quad = \quad 9$

$4 + 5 \quad \neq \quad 9$

30.

$4 + 7 \quad = \quad 10$

$4 + 7 \quad \neq \quad 10$

OA2-10 Using Doubles to Add (I)

◯ Write a doubles sentence.

1.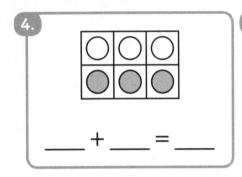

___4___ + ___4___ = ___8___

2.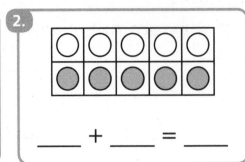

___ + ___ = ___

3.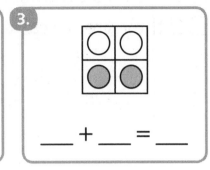

___ + ___ = ___

4.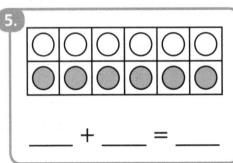

___ + ___ = ___

5.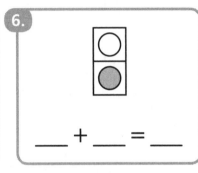

___ + ___ = ___

6.

___ + ___ = ___

◯ Write a doubles addition sentence.

7. Jake has 3 cats. Grace has 3 cats.

Altogether, they have ___3___ + ___3___ = ___6___ cats.

8. Tina has 2 books. Sun has 2 books.

Altogether, they have _____ + _____ = _____ books.

9. Ivan has 5 pencils. Emma has 5 pencils.

Altogether, they have _____ + _____ = _____ pencils.

10. Alex has 4 hats. Raj has 4 hats.

Altogether, they have _____ + _____ = _____ hats.

☐ Write an addition for each picture.
Are the two additions equal?
☐ If they are, circle =. If they are not, circle ≠.

11.

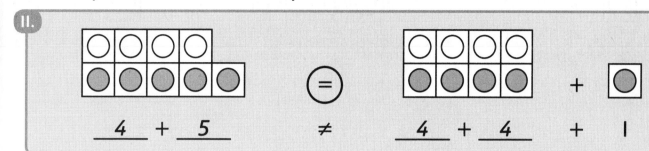

___4___ + ___5___   ≠   ___4___ + ___4___   +   1

12.

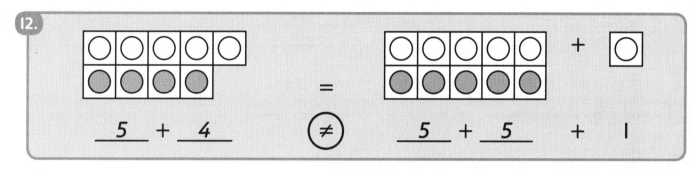

___5___ + ___4___   (≠)   ___5___ + ___5___   +   1

13.

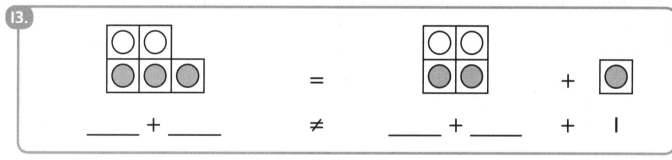

____ + ____   ≠   ____ + ____   +   1

14.

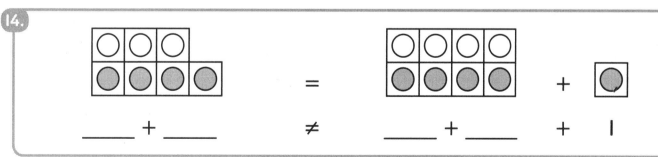

____ + ____   ≠   ____ + ____   +   1

15.

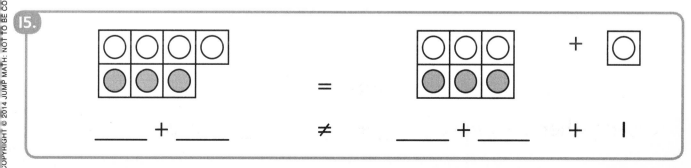

____ + ____   ≠   ____ + ____   +   1

◻ Circle the smaller number.

◻ Write the double of the smaller number plus 1.

16.

$$⑤ + 6$$

$$= \underline{\ 5\ } + \underline{\ 5\ } + \underline{\ 1\ }$$

17.

$$6 + ⑤$$

$$= \underline{\ 5\ } + \underline{\ 5\ } + \underline{\ 1\ }$$

18.

$$4 + 3$$

$$= \underline{\quad} + \underline{\quad} + \underline{\quad}$$

19.

$$3 + 4$$

$$= \underline{\quad} + \underline{\quad} + \underline{\quad}$$

20.

$$6 + 7$$

$$= \underline{\quad} + \underline{\quad} + \underline{\quad}$$

21.

$$7 + 6$$

$$= \underline{\quad} + \underline{\quad} + \underline{\quad}$$

◻ Write the double of the smaller number.

◻ Add the double.

◻ Add 1.

22.

$$4 + 5$$

$$= \underline{\ 4\ } + \underline{\ 4\ } + 1$$

$$= \underline{\ 8\ } + 1$$

$$= \underline{\ 9\ }$$

23.

$$5 + 4$$

$$= \underline{\quad} + \underline{\quad} + 1$$

$$= \underline{\quad} + 1$$

$$= \underline{\quad}$$

24.

$$2 + 3$$

$$= \underline{\quad} + \underline{\quad} + 1$$

$$= \underline{\quad} + 1$$

$$= \underline{\quad}$$

25.

$$3 + 2$$

$$= \underline{\quad} + \underline{\quad} + 1$$

$$= \underline{\quad} + 1$$

$$= \underline{\quad}$$

26.

$$5 + 6$$

$$= \underline{\quad} + \underline{\quad} + 1$$

$$= \underline{\quad} + 1$$

$$= \underline{\quad}$$

27.

$$6 + 5$$

$$= \underline{\quad} + \underline{\quad} + 1$$

$$= \underline{\quad} + 1$$

$$= \underline{\quad}$$

Operations and Algebraic Thinking 2-10

OA2-11 Using Doubles to Add (2)

☐ Write the number that is between.

1. 5, __6__ , 7

2. 1, _____ , 3

3. 6, _____ , 8

4. 7 and 9 __8__

5. 4 and 6 _____

6. 8 and 10 _____

☐ Write an addition for each picture.

7.
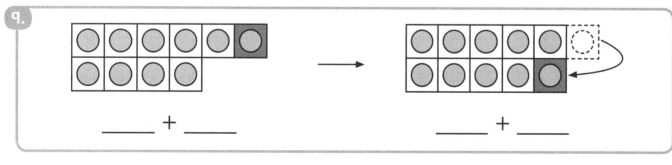

__3__ + __5__   __4__ + __4__

8.
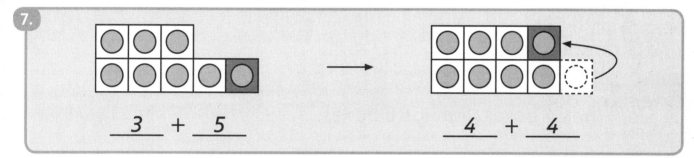

__4__ + __2__   __3__ + __3__

9.
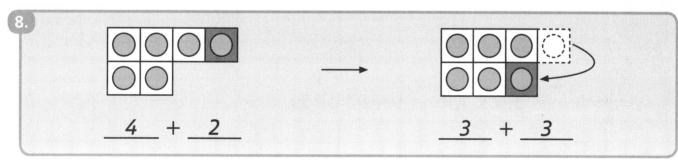

_____ + _____   _____ + _____

10.
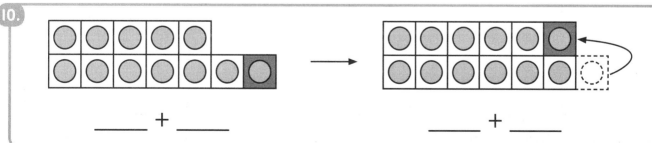

_____ + _____   _____ + _____

☐ Write the addition as a double.

11.

$3 + 5 \quad = \quad \underline{\ 4\ } + \underline{\ 4\ }$

$= \quad \underline{\ 8\ }$

12.

$1 + 3 \quad = \quad \underline{\qquad} + \underline{\qquad}$

$= \quad \underline{\qquad}$

13.

$6 + 8 \quad = \quad \underline{\qquad} + \underline{\qquad}$

$= \quad \underline{\qquad}$

14.

$5 + 7 \quad = \quad \underline{\qquad} + \underline{\qquad}$

$= \quad \underline{\qquad}$

☐ Use doubles to add.

15.

Jane has 4 boxes. Sal has 6 boxes.

They have _____ + _____ = _____ boxes in total.

16.

Amit sees 5 birds. Mona sees 3 birds.

They see _____ + _____ = _____ birds in total.

17.

Kim counts 5 cars. Tony counts 7 cars.

They count _____ + _____ = _____ cars in total.

18.

Vicky plants 1 tree. John plants 3 trees.

They plant _____ + _____ = _____ trees in total.

OA2-12 Distance from 0 on a Number Line

☐ Circle the correct number line.

1.

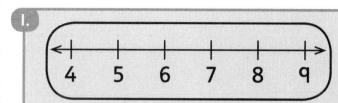

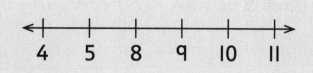

2.

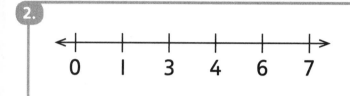

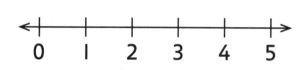

3.

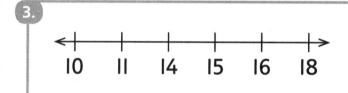

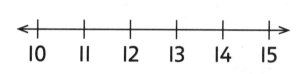

The frog starts at 0.

☐ Number each jump.

☐ Circle the distance from 0.

4.

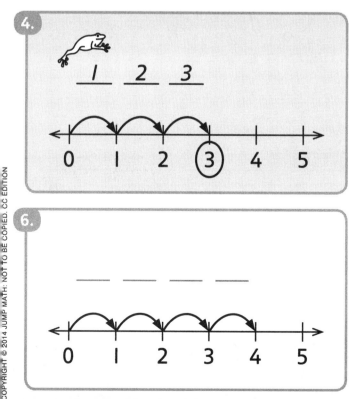

5.

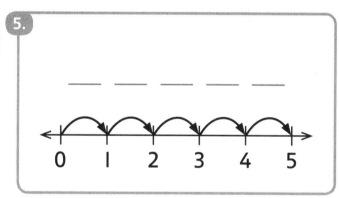

6.

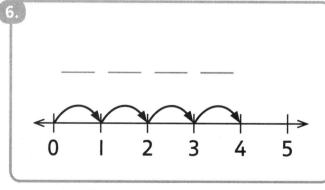

7.

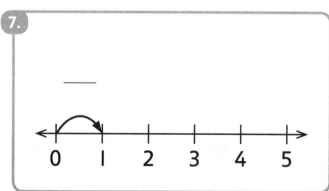

How many jumps away from 0 is the frog?

☐ Number each jump.

☐ Circle the distance from 0.

8.

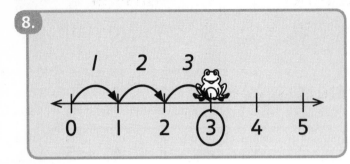

9.

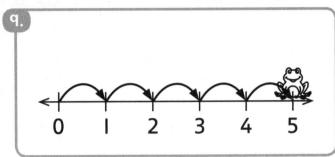

10.

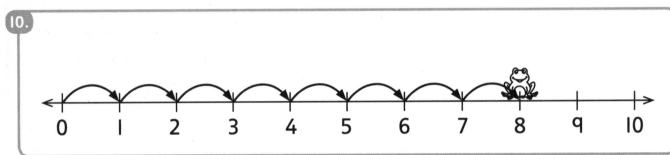

11.

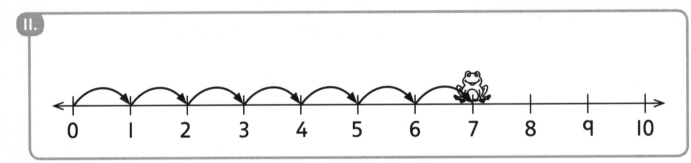

12.

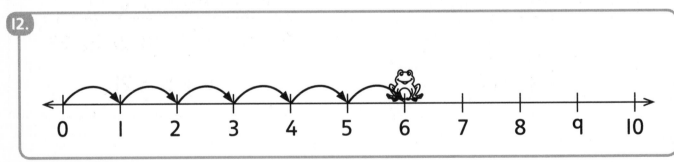

13.

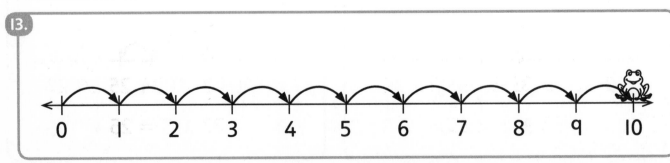

Operations and Algebraic Thinking 2-12

OA2-13 Addition Using a Number Line

3	+	2	=	5
first number		second number		answer

The frog starts at the first number.

☐ Circle the first number in the addition sentence.

☐ Draw a dot where the frog starts.

1.
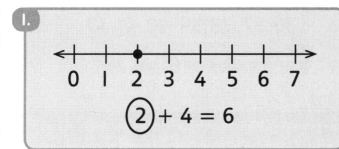

$$2 + 4 = 6$$

2.
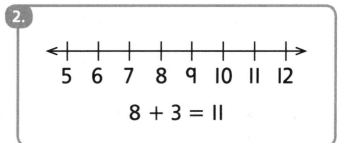

$$8 + 3 = 11$$

The frog makes the second number of jumps.

☐ Circle the second number in the addition sentence.

☐ Number each jump.

3.
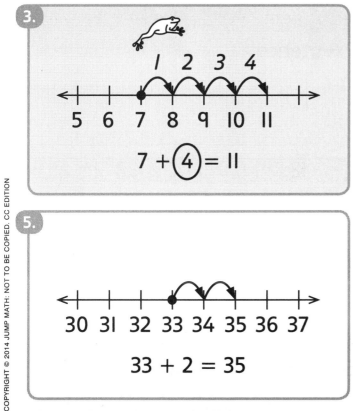

$$7 + 4 = 11$$

4.
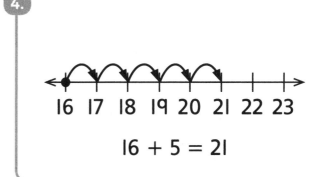

$$16 + 5 = 21$$

5.
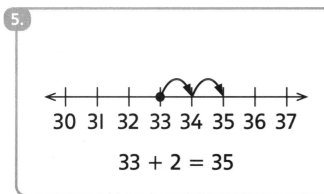

$$33 + 2 = 35$$

6.
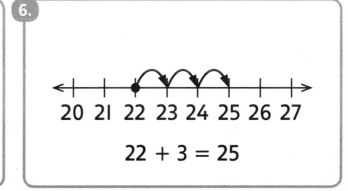

$$22 + 3 = 25$$

⬭ Trace the jumps.

7.

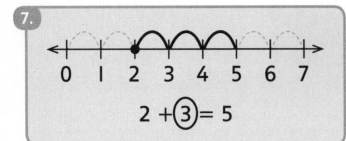

$2 + ③ = 5$

8.

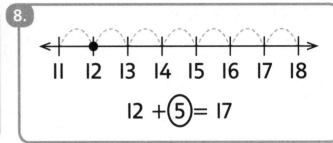

$12 + ⑤ = 17$

9.

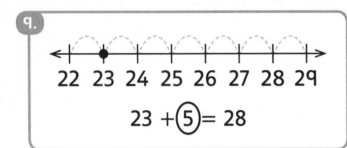

$23 + ⑤ = 28$

10.

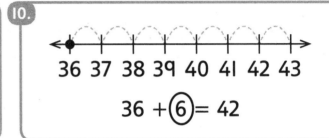

$36 + ⑥ = 42$

11.

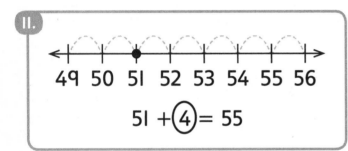

$51 + ④ = 55$

12.

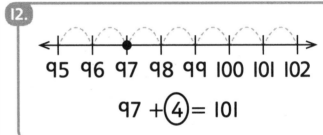

$97 + ④ = 101$

The frog stops at the answer.
⬭ Circle the answer in the addition sentence.
⬭ Draw a dot where the frog stops.

13.

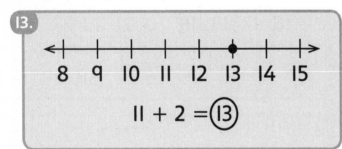

$11 + 2 = ⑬$

14.

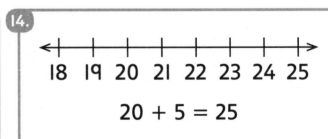

$20 + 5 = 25$

15.

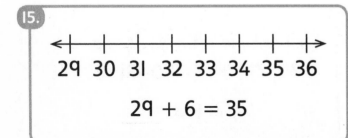

$29 + 6 = 35$

16.

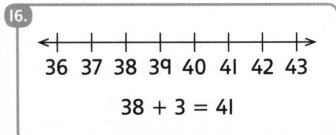

$38 + 3 = 41$

Use a number line to add.

☐ Trace the jumps.

☐ Add.

17.

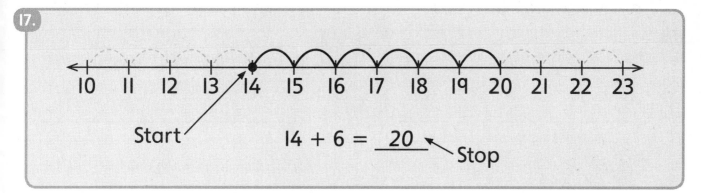

Start

14 + 6 = ___20___

Stop

18.

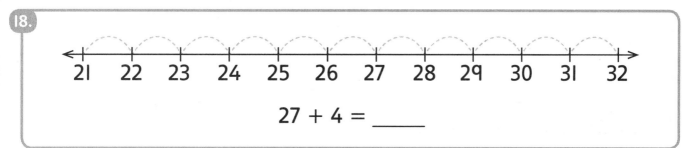

27 + 4 = _____

19.

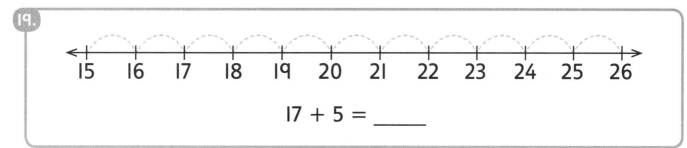

17 + 5 = _____

20.

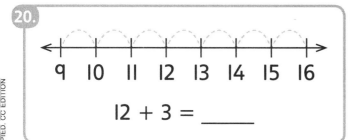

12 + 3 = _____

21.

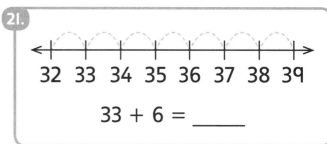

33 + 6 = _____

22.

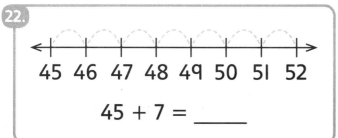

45 + 7 = _____

23.

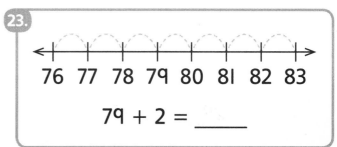

79 + 2 = _____

OA2-14 Making a Number Line to Add

The frog starts at the first number.
- Draw a dot where the frog starts.
- Write the number.

1.
$$2 + 3 = 5$$
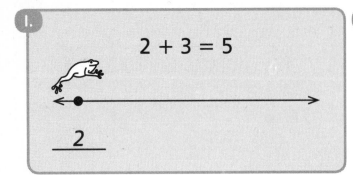
__2__

2.
$$8 + 7 = 15$$
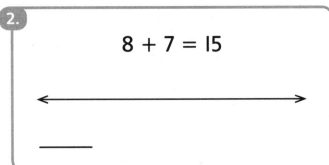
_____

3.
$$14 + 9 = 23$$
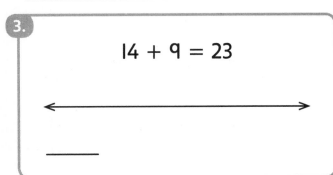
_____

4.
$$63 + 5 = 68$$
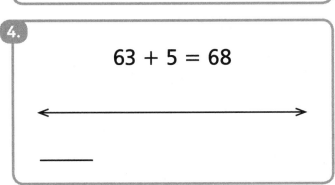
_____

- Draw the first jump.
- Write the number for the first jump.

5.
$$7 + 2 = 9$$
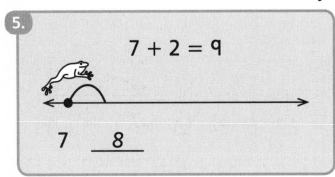
7   __8__

6.
$$14 + 1 = 15$$
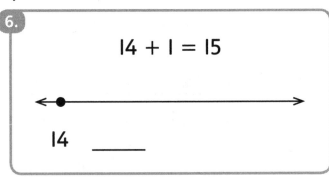
14   _____

7.
$$23 + 6 = 29$$
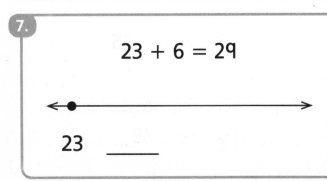
23   _____

8.
$$40 + 8 = 48$$
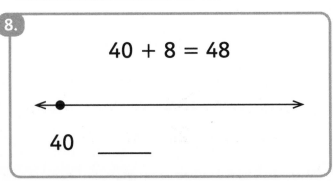
40   _____

Operations and Algebraic Thinking 2-14

☐ Write the number for each jump.
☐ Add.

9.

$8 + 4 = \underline{12}$

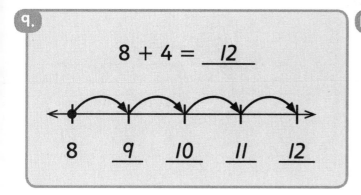

8   _9_   _10_   _11_   _12_

10.

$13 + 3 = \underline{}$

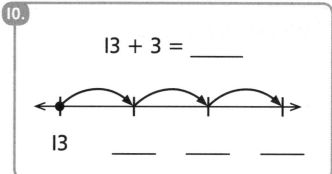

13   ___   ___   ___

11.

$22 + 5 = \underline{}$

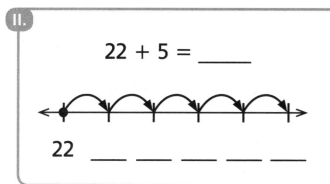

22   ___ ___ ___ ___ ___

12.

$31 + 2 = \underline{}$

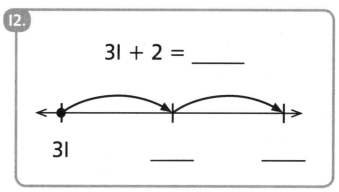

31   ___   ___

☐ Draw a tick mark where the frog starts.
☐ Draw a tick mark for each jump.

13.

$11 + 3 = 14$

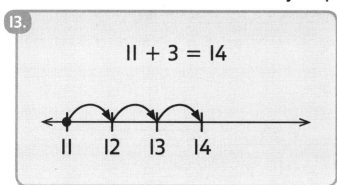

11   12   13   14

14.

$14 + 5 = 19$

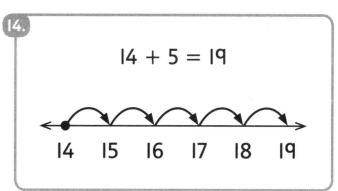

14   15   16   17   18   19

15.

$23 + 4 = 27$

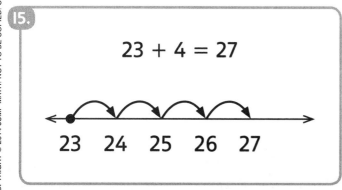

23   24   25   26   27

16.

$32 + 3 = 35$

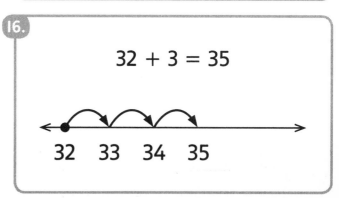

32   33   34   35

☐ Circle the number of jumps.
☐ Draw the jumps on the number line.

17.
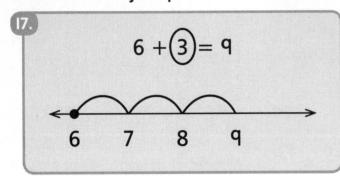
$6 + ③ = 9$

18.
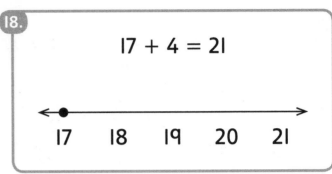
$17 + 4 = 21$

19.
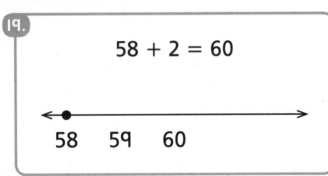
$58 + 2 = 60$

20.
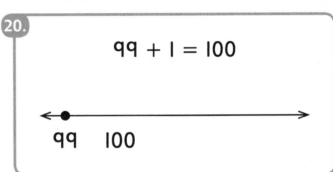
$99 + 1 = 100$

It is hard to draw a lot of jumps.
☐ Circle the addition sentence with fewer jumps.

21.
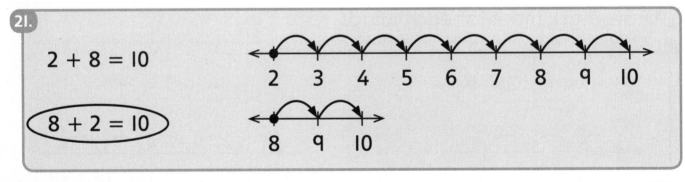
$2 + 8 = 10$

$⟨8 + 2 = 10⟩$

22.
$⟨10 + 4 = 14⟩$
$4 + 10 = 14$

23.
$6 + 17 = 23$
$17 + 6 = 23$

24.
$33 + 5 = 38$
$5 + 33 = 38$

25.
$2 + 49 = 51$
$49 + 2 = 51$

26.
$3 + 64 = 67$
$64 + 3 = 67$

27.
$98 + 1 = 99$
$1 + 98 = 99$

Operations and Algebraic Thinking 2-14

☐ Draw a number line.
☐ Use the number line to add.

28.

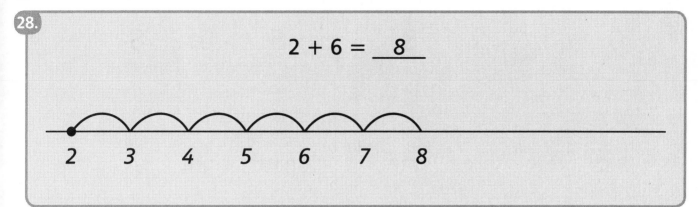

$$2 + 6 = \underline{8}$$

29.

$$3 + 7 = \underline{}$$

30.

$$11 + 5 = \underline{}$$

31.

$$20 + 4 = \underline{}$$

OA2-I5 Number Words from 0 to I0

☐ Match the numbers to the words.

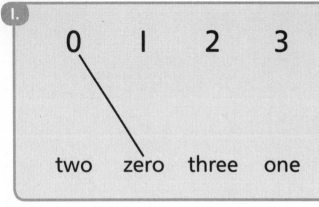

1.

0	1	2	3

two   zero   three   one

2.

5	6	8	9

nine   eight   six   five

3.

4	7	6	8	3	9

seven   three   four   nine   six   eight

4.

1	4	2	6	5	8

six   one   two   eight   four   five

5.

7	3	9	2	1	4

three   nine   four   seven   two   one

☐ Write the numbers above the number words.

6.

$$8 \qquad 1$$

Hanna has eight pencils and one eraser.

7.

Amit is nine years old and Sam is ten years old.

8.

Pam has seven crayons, two markers, and zero pens.

9.

Rob has five brothers and his sister has six brothers.

10.

Liz has three sisters and her brother has four sisters.

☐ Circle the number words.

11.

(t h r e e)(t w o)(n i n e)(s i x)(e i g h t)

12.

t e n s e v e n f i v e f o u r o n e t h r e e

OA2-16 Number Words from 0 to 20

☐ Underline the beginning letters that are the same.

1.
two   twelve

2.
six   sixteen

3.
three   thirteen

4.
four   fourteen

5.
eight   eighteen

6.
five   fifteen

☐ Circle the digits that are the same.

7.
② 1②

8.
6   16

9.
7   17

10.
9   19

11.
8   18

12.
3   13

☐ Underline and circle the same parts.

13.
three = ③
thirteen = 1③

14.
four = 4
fourteen = 14

15.
five = 5
fifteen = 15

16.
nine = 9
nineteen = 19

17.
seven = 7
seventeen = 17

18.
two = 2
twelve = 12

☐ Write the number.

19.
thirteen = _1_ _3_

20.
seventeen = __ __

21.
fifteen = __ __

22.
sixteen = __ __

23.
fourteen = __ __

24.
twelve = __ __

25.
nineteen = __ __

26.
eighteen = __ __

27.
eleven = __ __

☐ Match the word with the number.

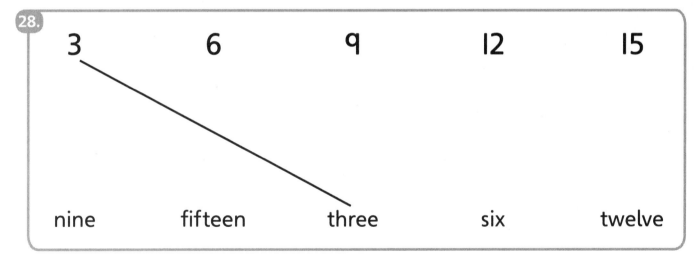

28.

| 3 | 6 | 9 | 12 | 15 |

| nine | fifteen | three | six | twelve |

29.

| 11 | 13 | 15 | 17 | 19 |

| fifteen | nineteen | thirteen | eleven | seventeen |

☐ Write the number above the number word.

30.
13
Clara is thirteen months old.

31.
Mike has twenty teeth.

32.
Sixteen friends played tag.

33.
Holidays start in eleven days.

34.
We played basketball for fifteen minutes.

35.
Will invited eighteen friends to his birthday party.

36. BONUS
Bianca's soccer team has twelve players.

There are seven girls and five boys.

37. BONUS
There are eight pears and twelve plums on the table.

38. BONUS
One week has seven days. Two days are on the weekend.

○ Answer the question using the number and the word.

39. What grade are you in? __2__ = ___two___

40. How many letters are in your first name? _____ = _____

41. How old are you? _____ = _____

42. How many pets do you have? _____ = _____

43. How many girls are in your class? _____ = _____

44. How many boys are in your class? _____ = _____

45. How many months are in a year? _____ = _____

46. How many blank lines (__) are on this page? _____ = _____

47. BONUS How many letters are black? _____ = _____

a **b c d** e **f g h** i **j k l m n** o **p q r s t** u **v w** x y **z**

OA2-17 Numbers from 10 to 19

1.

Josh has 10 apples.

Josh gets more apples.

10 + __1__ = __1__ __1__ apples   10 + __2__ = __1__ __2__ apples

☐ How many apples altogether? Add.

2.

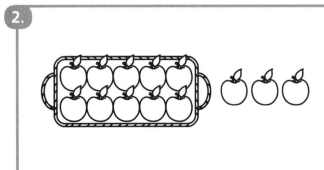

10 + ___ = __1__ ___ apples

3.

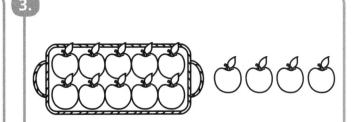

10 + ___ = ___ ___ apples

☐ Add.

4.

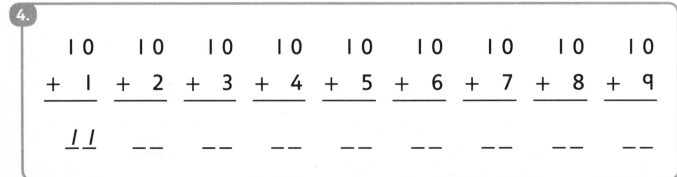

10	10	10	10	10	10	10	10	10
+ 1	+ 2	+ 3	+ 4	+ 5	+ 6	+ 7	+ 8	+ 9
1 1	— —	— —	— —	— —	— —	— —	— —	— —

☐ How many?

5.

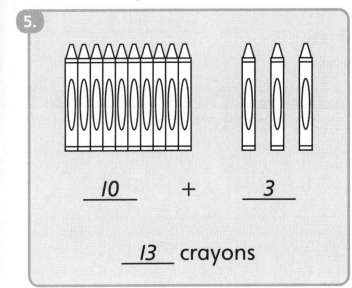

___10___ + ___3___

___13___ crayons

6.

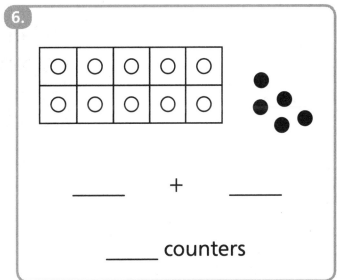

_____ + _____

_____ counters

7.

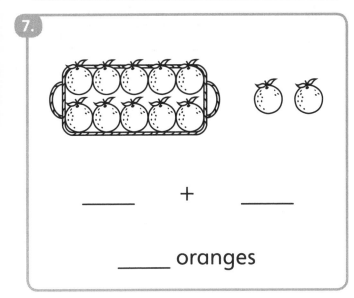

_____ + _____

_____ oranges

8.

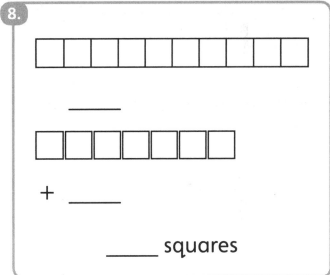

_____

+ _____

_____ squares

9.

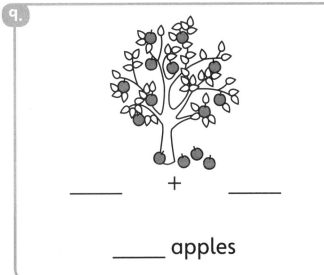

_____ + _____

_____ apples

10.

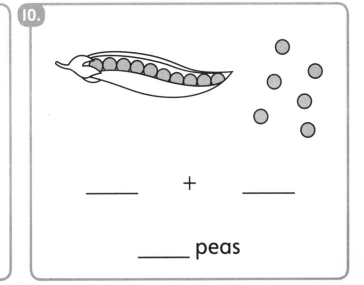

_____ + _____

_____ peas

OA2-18 Using 10 to Add

☐ Use the group of 10 to help you add.

1.
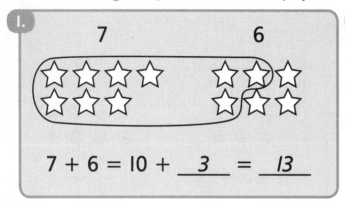

7   6

7 + 6 = 10 + __3__ = __13__

2.
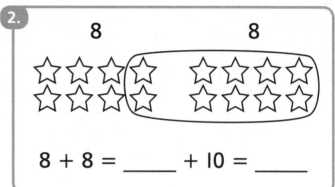

8   8

8 + 8 = ____ + 10 = ____

3.
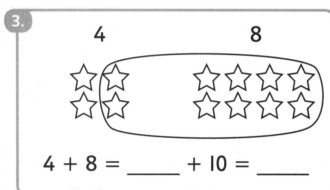

4   8

4 + 8 = ____ + 10 = ____

4.
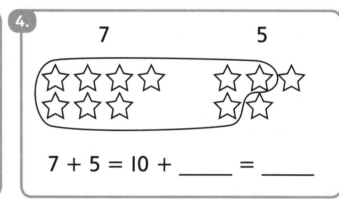

7   5

7 + 5 = 10 + ____ = ____

5.
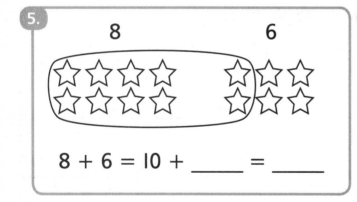

8   6

8 + 6 = 10 + ____ = ____

6.
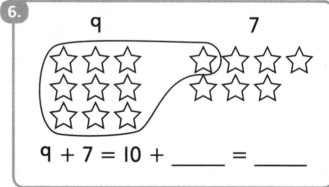

9   7

9 + 7 = 10 + ____ = ____

☐ Sara groups 10 in two ways. Does she get the same answer?

7.
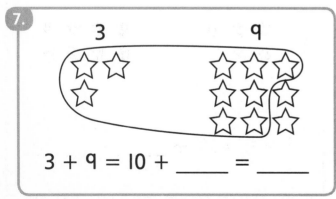

3   9

3 + 9 = 10 + ____ = ____

8.
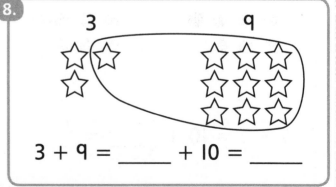

3   9

3 + 9 = ____ + 10 = ____

○ Circle a group of 10.
○ Use 10 to add.

9.

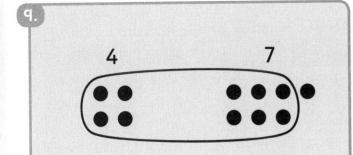

4   7

$4 + 7 = 10 + \underline{\quad 1 \quad} = \underline{\quad 11 \quad}$

10.

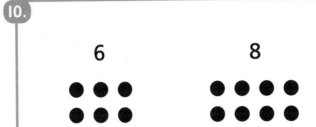

6   8

$6 + 8 = \underline{\quad\quad} + 10 = \underline{\quad\quad}$

11.

4   9

$4 + 9 = \underline{\quad\quad} + 10 = \underline{\quad\quad}$

12.

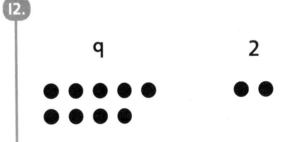

9   2

$9 + 2 = 10 + \underline{\quad\quad} = \underline{\quad\quad}$

13.

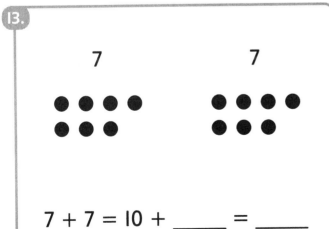

7   7

$7 + 7 = 10 + \underline{\quad\quad} = \underline{\quad\quad}$

14.

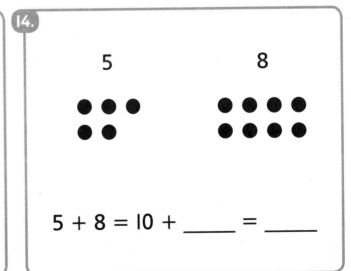

5   8

$5 + 8 = 10 + \underline{\quad\quad} = \underline{\quad\quad}$

Operations and Algebraic Thinking 2-18

OA2-19 Pairs of Numbers That Add to 10

☐ Count on to 10.

1.

6 | 7 | 8 | 9 | 10

$6 + \underline{\ 4\ } = 10$

5 | 6 | 7 | 8 | 9 | 10

$5 + \underline{\ 5\ } = 10$

2. $5 + \underline{\hspace{1cm}} = 10$

3. $8 + \underline{\hspace{1cm}} = 10$

4. $0 + \underline{\hspace{1cm}} = 10$

5. $4 + \underline{\hspace{1cm}} = 10$

6. $7 + \underline{\hspace{1cm}} = 10$

7. $9 + \underline{\hspace{1cm}} = 10$

☐ Write an addition sentence for the picture.

8.

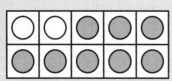

$\underline{\ 2\ } + \underline{\ 8\ } = 10$

9.

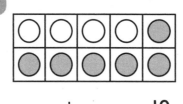

$\underline{\hspace{0.8cm}} + \underline{\hspace{0.8cm}} = 10$

10.

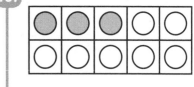

$\underline{\hspace{0.8cm}} + \underline{\hspace{0.8cm}} = 10$

11.

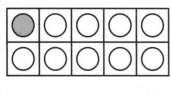

$\underline{\hspace{0.8cm}} + \underline{\hspace{0.8cm}} = 10$

12.

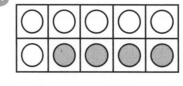

$\underline{\hspace{0.8cm}} + \underline{\hspace{0.8cm}} = 10$

13.

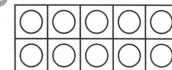

$\underline{\hspace{0.8cm}} + \underline{\hspace{0.8cm}} = 10$

Operations and Algebraic Thinking 2-19

◯ Draw circles in the empty boxes.
◯ Complete the addition sentence.

14.
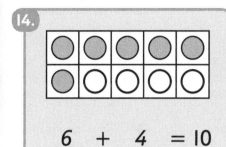

__6__ + __4__ = 10

15.
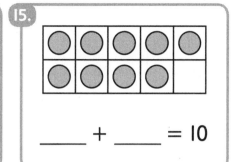

____ + ____ = 10

16.
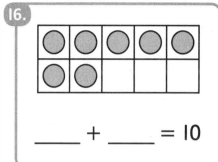

____ + ____ = 10

17.
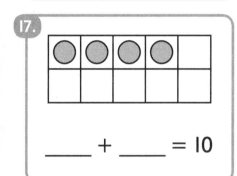

____ + ____ = 10

18.
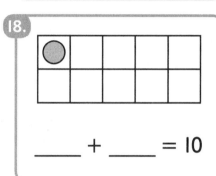

____ + ____ = 10

19.
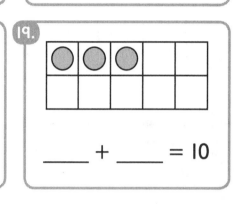

____ + ____ = 10

◯ Write two addition sentences for the picture.

20.
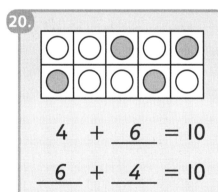

4 + __6__ = 10

__6__ + __4__ = 10

21.
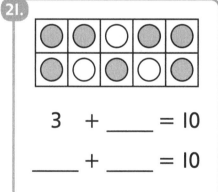

3 + ____ = 10

____ + ____ = 10

22.
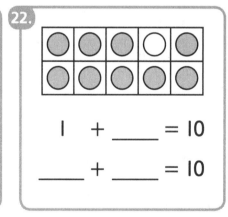

1 + ____ = 10

____ + ____ = 10

23.
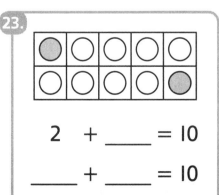

2 + ____ = 10

____ + ____ = 10

24.
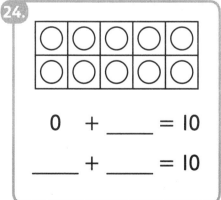

0 + ____ = 10

____ + ____ = 10

25.
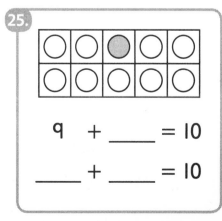

9 + ____ = 10

____ + ____ = 10

☐ Underline the number that makes 10.
☐ Write the addition sentence.

26.
0   1   ②   3   4   5   6   7   <u>8</u>   9   10   <u>2</u> + <u>8</u> = 10

27.
0   ①   2   3   4   5   6   7   8   9   10   ___ + ___ = 10

28.
0   1   2   3   4   5   ⑥   7   8   9   10   ___ + ___ = 10

29.
⓪   1   2   3   4   5   6   7   8   9   10   ___ + ___ = 10

30.
0   1   2   3   4   5   6   ⑦   8   9   10   ___ + ___ = 10

31.
⑧   4   9   0   10   5   6   3   <u>2</u>   7   1   <u>8</u> + <u>2</u> = 10

32.
④   1   5   0   7   9   8   2   10   3   6   ___ + ___ = 10

33.
③   10   6   1   7   0   8   9   4   5   2   ___ + ___ = 10

34.
⑨   5   10   1   2   0   3   4   6   8   7   ___ + ___ = 10

35.
⑩   0   6   5   2   7   1   3   4   8   9   ___ + ___ = 10

Operations and Algebraic Thinking 2-19

OA2-20 Addition That Makes More Than 10 (1)

☐ Underline the blocks needed to make 10.

☐ Write the number.

1.

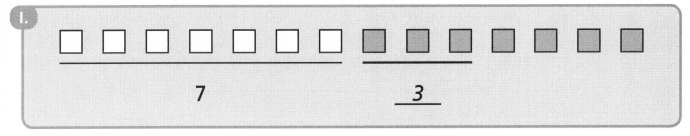

7   _3_

2.

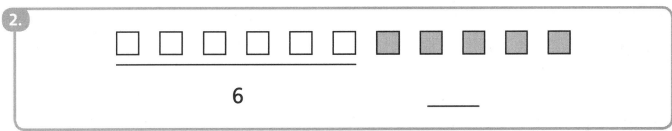

6   ____

3.

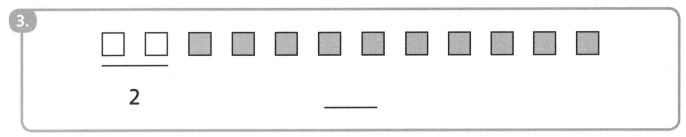

2   ____

4.

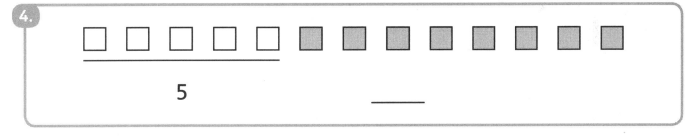

5   ____

5.

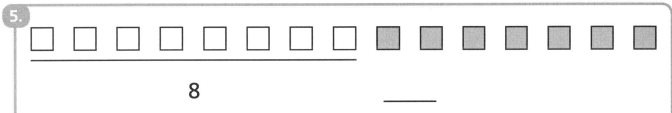

8   ____

6.

4   ____

☐ Underline the blocks needed to make 10.
☐ Circle the rest.
☐ Write the numbers.

7.

4   6   2

8.

5   ___   ___

9.

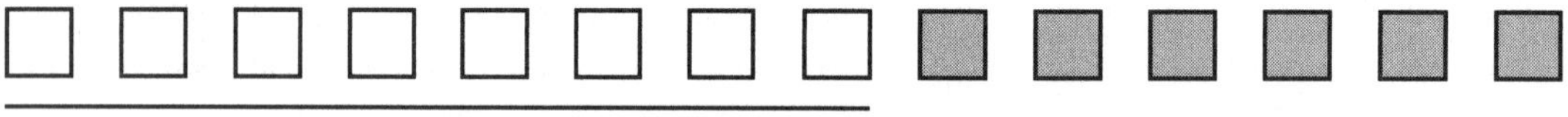

8   ___   ___

10.

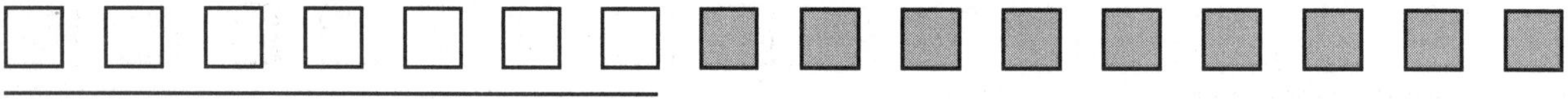

7   ___   ___

11.

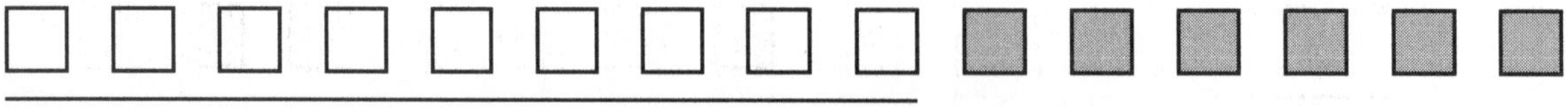

9   ___   ___

Operations and Algebraic Thinking 2-20

How many blocks altogether?
◯ How many blocks make 10? How many are left?
◯ Use 10 to add.

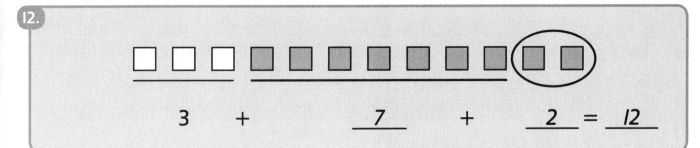

12.

$$3 \quad + \quad \underline{7} \quad + \quad \underline{2} = \underline{12}$$

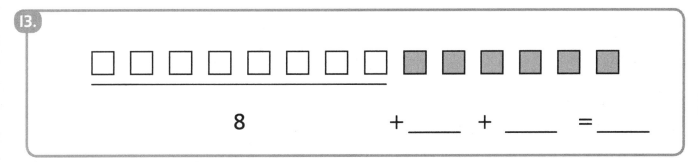

13.

$$8 \qquad +\underline{} + \underline{} = \underline{}$$

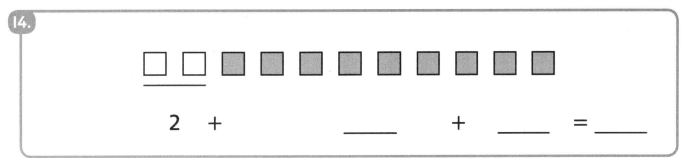

14.

$$2 \quad + \qquad \underline{} \quad + \quad \underline{} = \underline{}$$

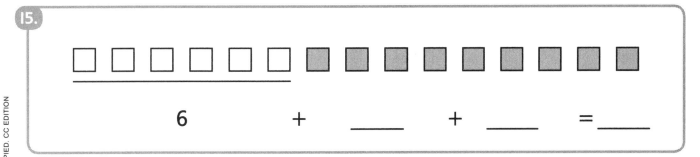

15.

$$6 \qquad + \quad \underline{} \quad + \quad \underline{} = \underline{}$$

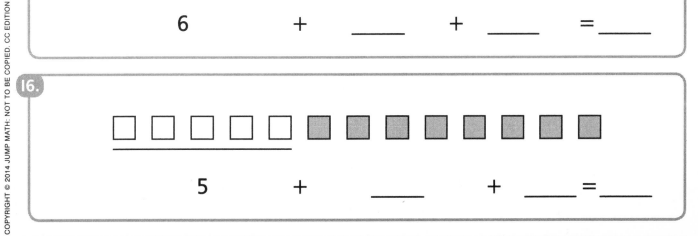

16.

$$5 \qquad + \qquad \underline{} \quad + \quad \underline{} = \underline{}$$

Operations and Algebraic Thinking 2-20

OA2-21 Addition That Makes More Than 10 (2)

☐ Add the white blocks and the gray blocks.
☐ Count all the blocks.

1.

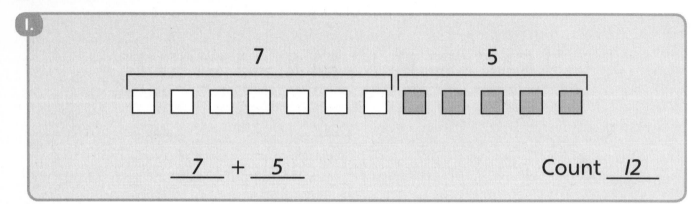

7   5

_7_ + _5_   Count _12_

2.

5   9

____ + ____   Count ____

3.

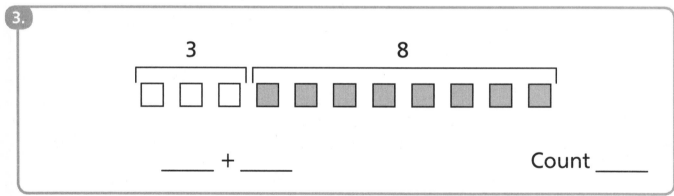

3   8

____ + ____   Count ____

4.

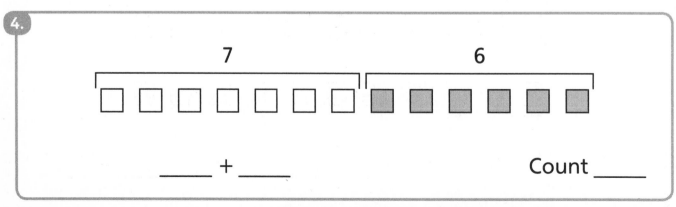

7   6

____ + ____   Count ____

☐ Underline the blocks needed to make 10.
☐ Circle the rest.
☐ Add.

5.

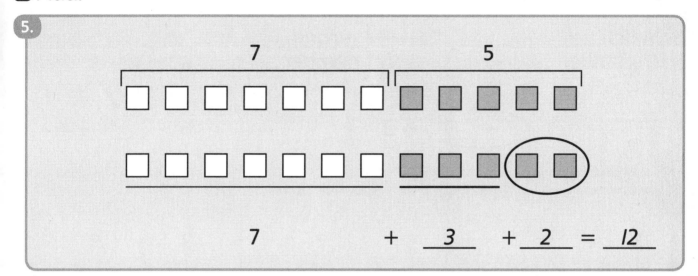

7   + _3_   + _2_ = _12_

6.

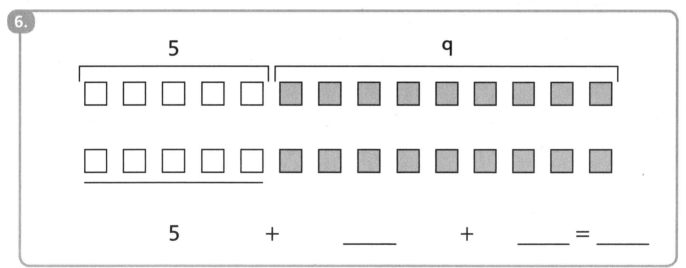

5   + _____   + _____ = _____

7.

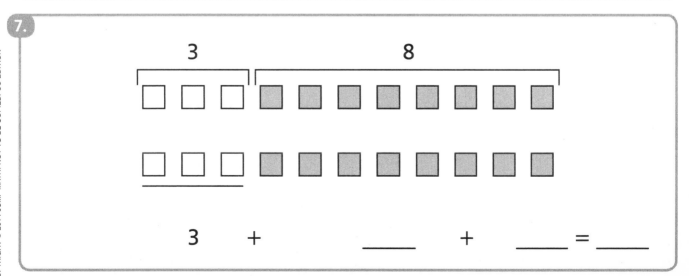

3   + _____   + _____ = _____

How much is needed from the second number to make 10?

⬜ Write the number.

⬜ Circle the numbers that add to 10.

8.

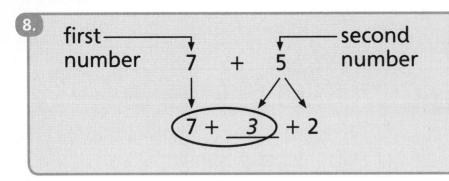

first number → 7 + 5 ← second number

$\left(7 + \underline{3}\right) + 2$

9.

$6 \quad + \quad 8$

$6 + \underline{} + 4$

10.

$5 \quad + \quad 11$

$5 + \underline{} + 6$

11.

$9 \quad + \quad 7$

$9 + \underline{} + 6$

12.

$3 \quad + \quad 9$

$3 + \underline{} + 2$

How much is needed from the second number to make 10?

How much is left?

⬜ Write the numbers.

13.

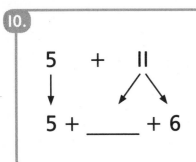

$8 \quad + \quad 3$

$8 + \underline{2} + \underline{1}$

14.

$5 \quad + \quad 7$

$5 + \underline{} + \underline{}$

15.

$9 \quad + \quad 4$

$9 + \underline{} + \underline{}$

16.

$7 \quad + \quad 6$

$7 + \underline{} + \underline{}$

17.

$8 \quad + \quad 8$

$8 + \underline{} + \underline{}$

18.

$7 \quad + \quad 9$

$7 + \underline{} + \underline{}$

Operations and Algebraic Thinking 2-21

OA2-22 Using Pictures to Subtract

☐ Use the picture to subtract.

1.

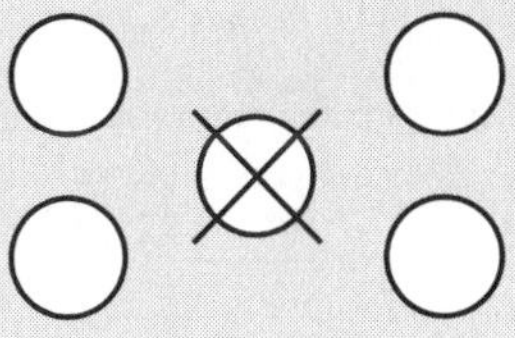

5 − 1 = 4

2.

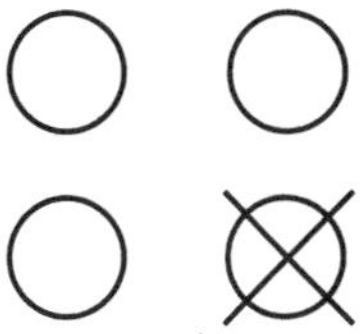

4 − 1 = ____

3.

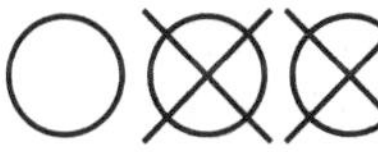

3 − 2 = ____

4.

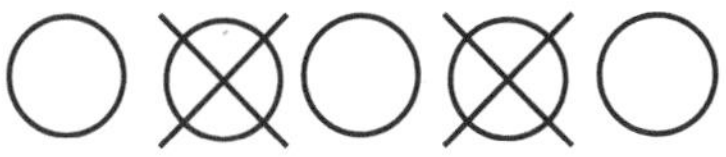

5 − 2 = ____

5.

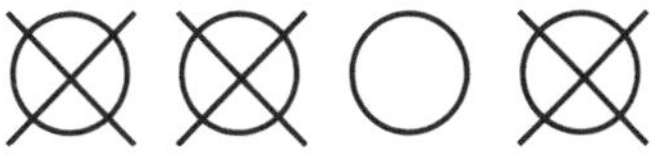

4 − 3 = ____

6.

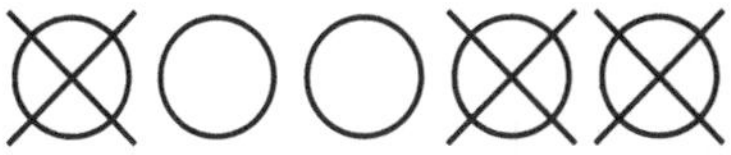

5 − 3 = ____

7.

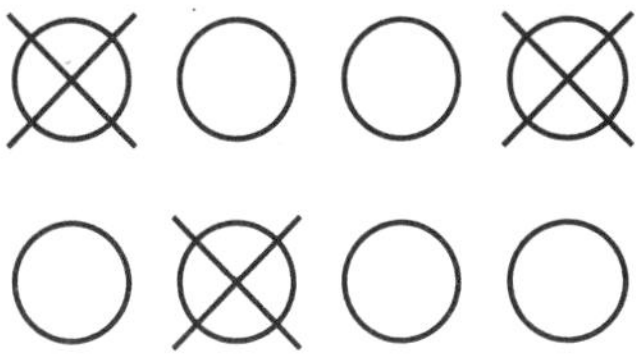

8 − 3 = ____

8.

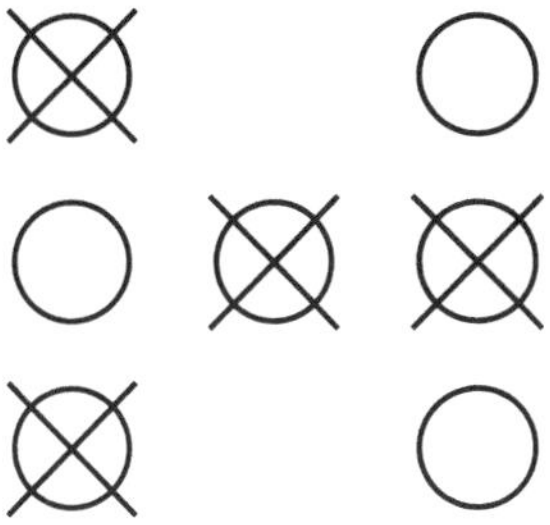

7 − 4 = ____

9.

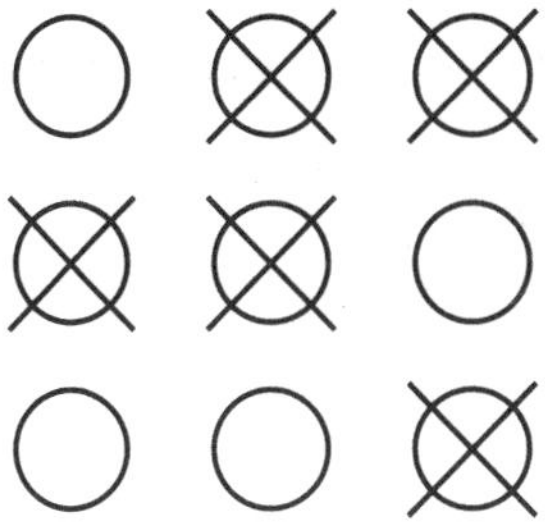

9 − 5 = ____

☐ Write a subtraction sentence for the picture.

10.

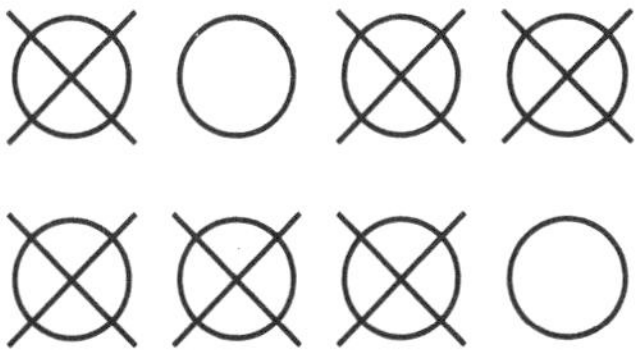

____ − ____ = ____

11.

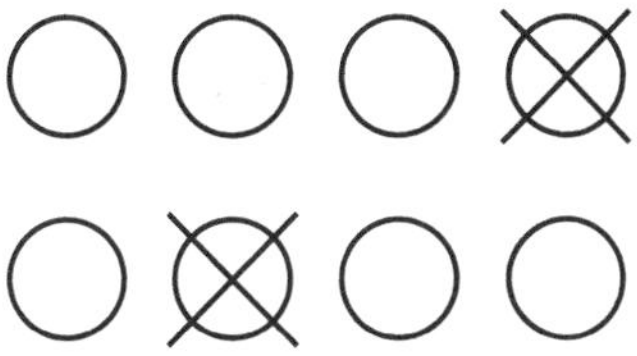

____ − ____ = ____

12.

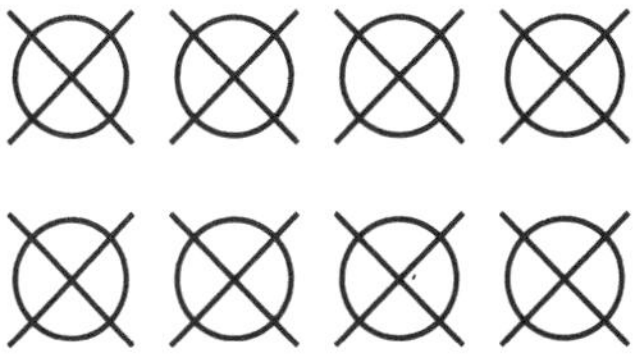

____ − ____ = ____

Use the picture to subtract.

13.

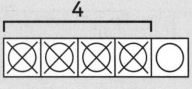

$5 - 4 = \underline{\quad 1 \quad}$

$5 - 1 = \underline{\quad 4 \quad}$

14.

$7 - 5 = \underline{\qquad}$

$7 - 2 = \underline{\qquad}$

15.

$6 - 6 = \underline{\qquad}$

$6 - 0 = \underline{\qquad}$

Use the picture to subtract the gray circles.

16.

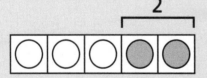

$5 - \underline{\quad 2 \quad} = \underline{\quad 3 \quad}$

17.

$5 - \underline{\qquad} = \underline{\qquad}$

18.

$6 - \underline{\qquad} = \underline{\qquad}$

19.

$6 - \underline{\qquad} = \underline{\qquad}$

☐ Write two subtraction sentences.

20.

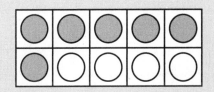

10 − __6__ = __4__

10 − __4__ = __6__

21.

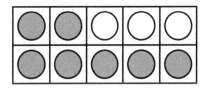

10 − ____ = ____

10 − ____ = ____

22.

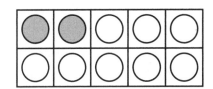

10 − ____ = ____

10 − ____ = ____

23.

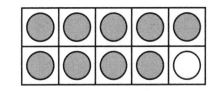

____ − ____ = ____

____ − ____ = ____

24.

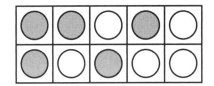

____ − ____ = ____

____ − ____ = ____

25.

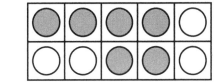

____ − ____ = ____

____ − ____ = ____

26.

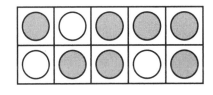

____ − ____ = ____

____ − ____ = ____

27.

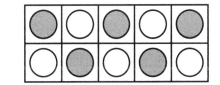

____ − ____ = ____

____ − ____ = ____

Operations and Algebraic Thinking 2-22

5	–	2	=	3
first number		second number		last number

The frog starts at the first number.

☐ Circle the first number in the subtraction sentence.

☐ Draw a dot where the frog starts.

1.

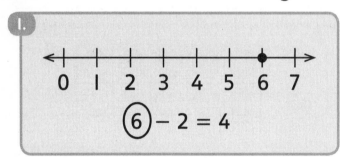

$6 - 2 = 4$

2.

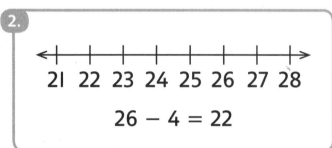

$26 - 4 = 22$

The frog makes the second number of jumps.

☐ Circle the second number in the subtraction sentence.

☐ Number each jump.

3.

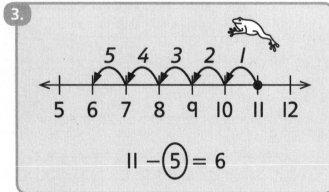

$11 - 5 = 6$

4.

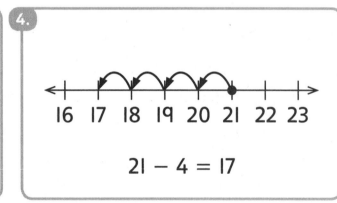

$21 - 4 = 17$

5.

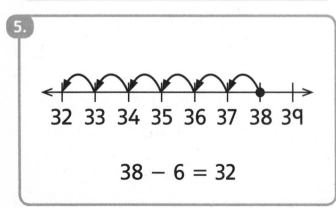

$38 - 6 = 32$

6.

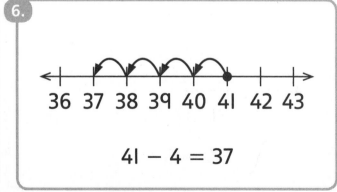

$41 - 4 = 37$

Trace the jumps.

Subtract.

7.
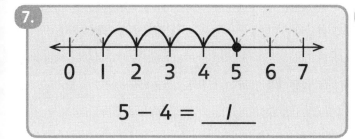

$5 - 4 =$ ___1___

8.
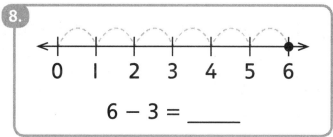

$6 - 3 =$ _____

9.
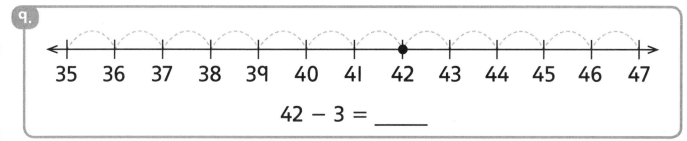

$42 - 3 =$ _____

10.
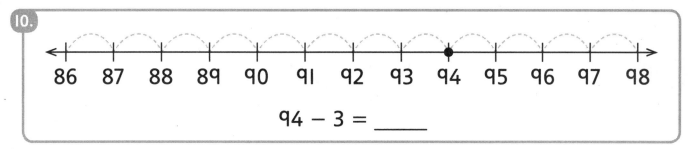

$94 - 3 =$ _____

The frog stops at the last number.

Circle the last number in the subtraction sentence.

Draw a dot where the frog stops.

11.
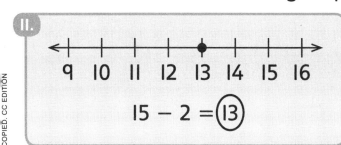

$15 - 2 = \boxed{13}$

12.
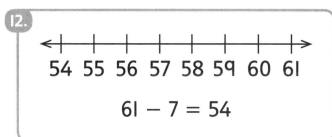

$61 - 7 = 54$

13.
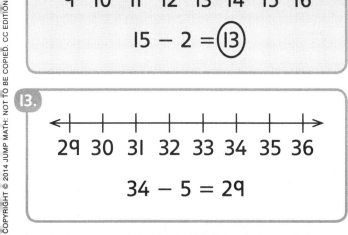

$34 - 5 = 29$

14.
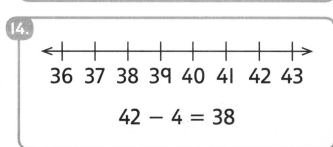

$42 - 4 = 38$

⬭ Use a number line to subtract.

15.
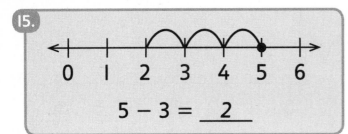

$5 - 3 = \underline{\ 2\ }$

16.
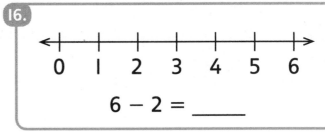

$6 - 2 = \underline{\hspace{1cm}}$

17.
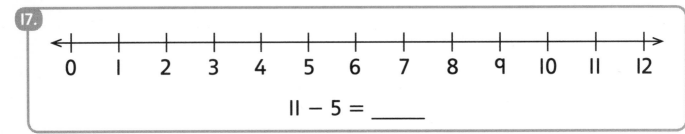

$11 - 5 = \underline{\hspace{1cm}}$

18.
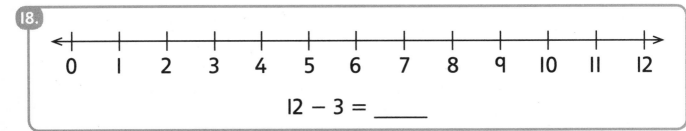

$12 - 3 = \underline{\hspace{1cm}}$

19.
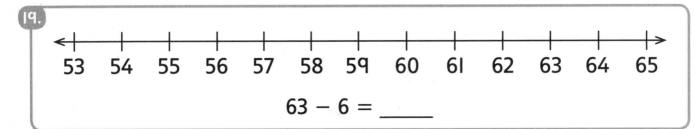

$63 - 6 = \underline{\hspace{1cm}}$

20.
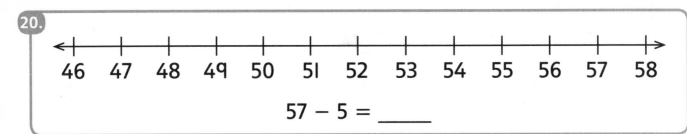

$57 - 5 = \underline{\hspace{1cm}}$

21.
Make your own.

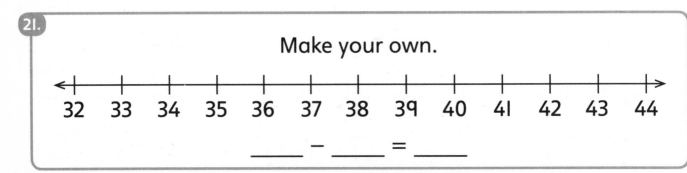

$\underline{\hspace{1cm}} - \underline{\hspace{1cm}} = \underline{\hspace{1cm}}$

Operations and Algebraic Thinking 2-23

OA2-24 Making a Number Line to Subtract

The frog starts at the first number.
- ◯ Draw a dot where the frog starts.
- ◯ Write the number.

1.
$$37 - 2 = 35$$
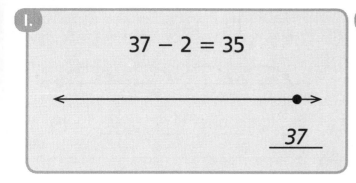
37

2.
$$25 - 3 = 22$$
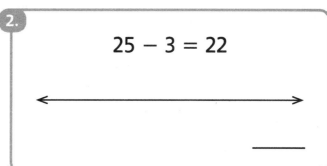
_____

3.
$$48 - 4 = 44$$
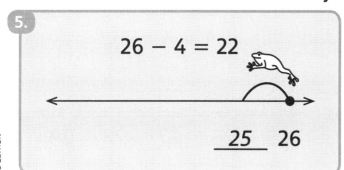
_____

4.
$$20 - 2 = 18$$
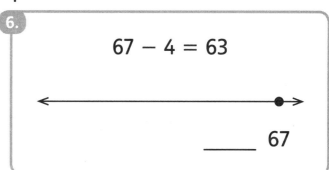
_____

- ◯ Draw the first jump.
- ◯ Write the number for the first jump.

5.
$$26 - 4 = 22$$
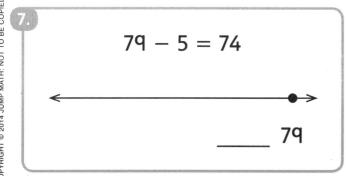
25   26

6.
$$67 - 4 = 63$$
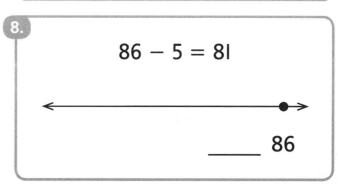
_____ 67

7.
$$79 - 5 = 74$$
_____ 79

8.
$$86 - 5 = 81$$
_____ 86

☐ Draw all the tick marks.
☐ Write the number for each jump.
☐ Subtract.

9.
$8 - 3 = \underline{\ 5\ }$

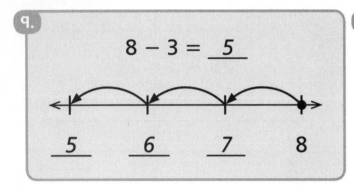

$\underline{\ 5\ }$ $\quad$ $\underline{\ 6\ }$ $\quad$ $\underline{\ 7\ }$ $\quad$ 8

10.
$19 - 4 = \underline{\quad}$

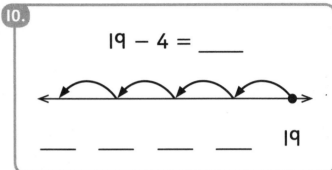

$\underline{\quad}$ $\underline{\quad}$ $\underline{\quad}$ $\underline{\quad}$ $\quad$ 19

11.
$33 - 2 = \underline{\quad}$

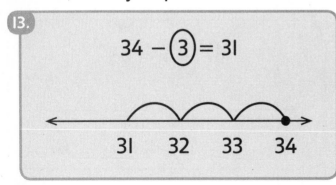

$\underline{\quad}$ $\underline{\quad}$ 33

12.
$40 - 5 = \underline{\quad}$

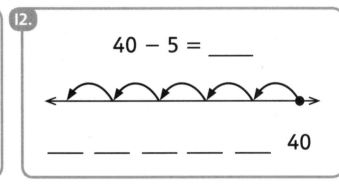

$\underline{\quad}$ $\underline{\quad}$ $\underline{\quad}$ $\underline{\quad}$ $\underline{\quad}$ 40

☐ Circle the number of jumps.
☐ Draw the jumps on the number line.

13.
$34 - ③ = 31$

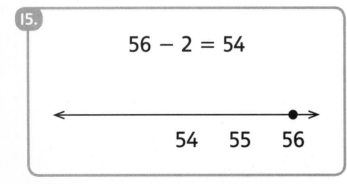

31 $\quad$ 32 $\quad$ 33 $\quad$ 34

14.
$29 - 4 = 25$

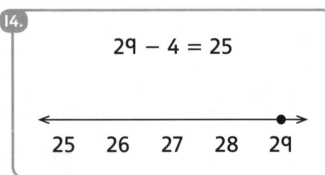

25 $\quad$ 26 $\quad$ 27 $\quad$ 28 $\quad$ 29

15.
$56 - 2 = 54$

54 $\quad$ 55 $\quad$ 56

16.
$67 - 3 = 64$

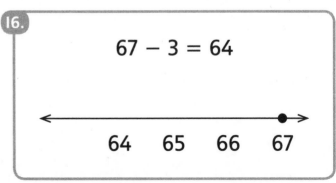

64 $\quad$ 65 $\quad$ 66 $\quad$ 67

☐ Draw a number line.
☐ Use the number line to subtract.

17.

$$8 - 5 = \underline{\ 3\ }$$

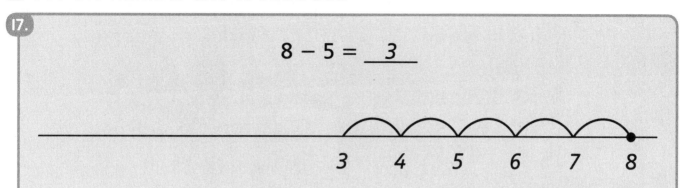

| 3 | 4 | 5 | 6 | 7 | 8 |

18.

$$39 - 4 = \underline{\qquad}$$

19.

$$37 - 7 = \underline{\qquad}$$

20.

$$45 - 6 = \underline{\qquad}$$

OA2-25 First Word Problems—Subtraction

☐ Subtract.

1.

5 apples in a tree

Jack eats 2 apples.
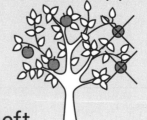

5 − 2 = _3_

There are _3_ apples left.

2.

8 birds

3 birds fly away.

8 − 3 = ____

There are ____ birds left.

3.

10 balloons

4 balloons break.

10 − 4 = ____

There are ____ balloons left.

4.

6 ducks

1 duck swims away.

6 − 1 = ____

There are ____ ducks left.

5.

5 windows

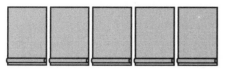

2 blinds are open.

5 − 2 = _____

_____ blinds are closed.

6.

4 books

Kate opens I book.

4 − 1 = _____

_____ books are closed.

7.

7 new crayons

David uses 2 crayons.

7 − 2 = _____

_____ crayons are not used.

8.

8 lights

5 lights are on.

8 − 5 = _____

_____ lights are off.

OA2-26 Add Two Ways Using a Number Line

The frog starts jumping from the dot.

☐ Draw the jumps.

1.

5 + 2

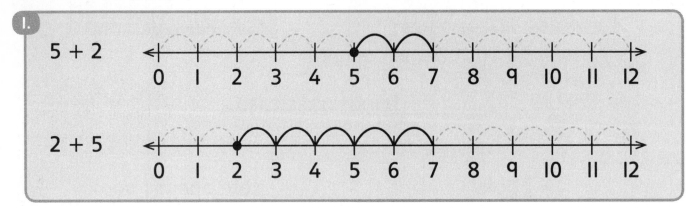

2 + 5

2.

3 + 8

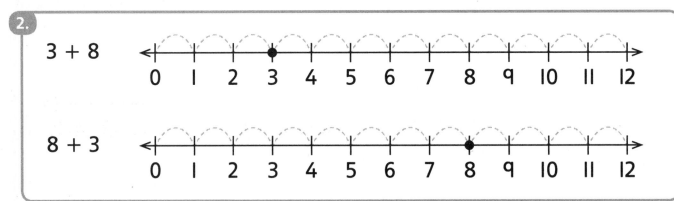

8 + 3

3.

7 + 4

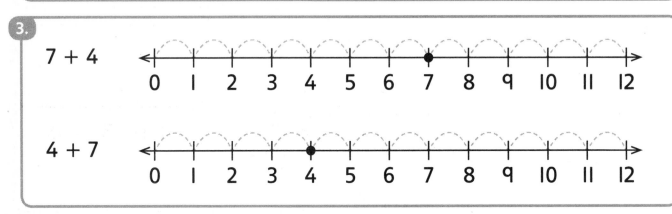

4 + 7

4.

1 + 11

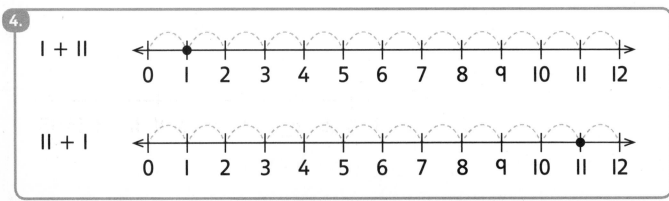

11 + 1

Operations and Algebraic Thinking 2-26

It is easier to draw the addend with fewer jumps.

◯ Draw jumps to add on the **smaller** addend.

◯ Add.

5.

$9 + 5 = \underline{\ 14\ }$

$5 + 9 = \underline{\hphantom{14}}$

6.

$2 + 11 = \underline{\hphantom{14}}$

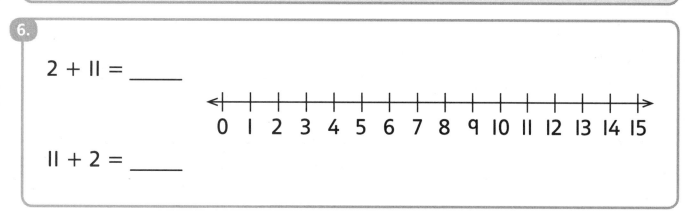

$11 + 2 = \underline{\hphantom{14}}$

7.

$10 + 3 = \underline{\hphantom{14}}$

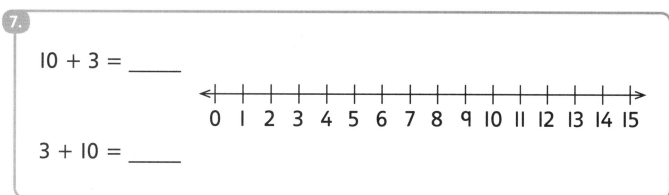

$3 + 10 = \underline{\hphantom{14}}$

8.

$8 + 4 = \underline{\hphantom{14}}$

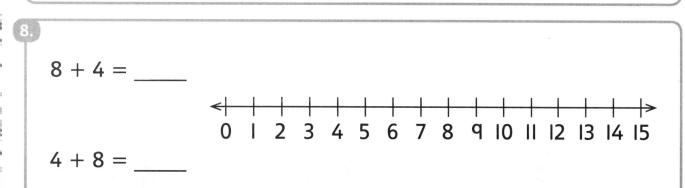

$4 + 8 = \underline{\hphantom{14}}$

OA2-27 Add Two Ways by Counting On

☐ Circle the addend you add on.

☐ Trace the correct number of blanks.

1.

$2 + ⑤$ __ __ __ __ __

$5 + ②$ __ __

2.

$6 + 3$

$3 + 6$

3.

$1 + 7$

$7 + 1$

☐ Trace the blanks for the addend you add on.

☐ Add by counting on.

4.

$4 + 8 = \underline{12}$ 4 _5_ _6_ _7_ _8_ _9_ _10_ _11_ _12_

$8 + 4 = \underline{12}$ 8 _9_ _10_ _11_ _12_

5.

$7 + 3 = \underline{}$ 7

$3 + 7 = \underline{}$ 3

6.

$2 + 9 = \underline{}$ 2

$9 + 2 = \underline{}$ 9

Operations and Algebraic Thinking 2-27

It is easier to add by counting on from the larger addend.

◯ Circle the addition that is easier to add.

7.

(14 + 3) — — —

3 + 14 — — — — — — — — — — — — —

8.

5 + 12 — — — — — — — — — — — — —

12 + 5 — — — — —

◯ Circle the addend that is easier to count from.

9. 6 +(18)   **10.** 23 + 4   **11.** 9 + 43   **12.** 76 + 8   **13.** 2 + 97

◯ Circle the addend that is easier to count from.
◯ Count on from that addend.

14.

4 +(9)= _13_   ⟨9⟩ _10_ _11_ _12_ _13_ ___ ___

15.

11 + 2 = ___   ⬡ ___ ___ ___ ___ ___ ___

16.

3 + 27 = ___   ⬡ ___ ___ ___ ___ ___ ___

17.

49 + 4 ___   ⬡ ___ ___ ___ ___ ___ ___

18.

5 + 85 ___   ⬡ ___ ___ ___ ___ ___ ___

OA2-28 Blocks in a Bag

Some blocks are in the bag.

☐ Circle the table where you know the exact number of blocks.

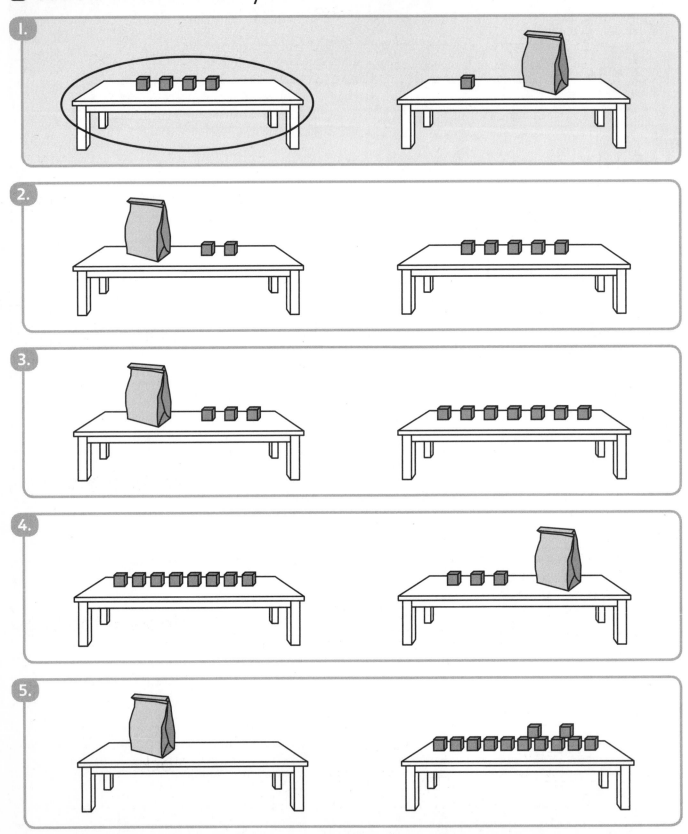

The two tables have the same number of blocks.
☐ How many blocks are in the bag?

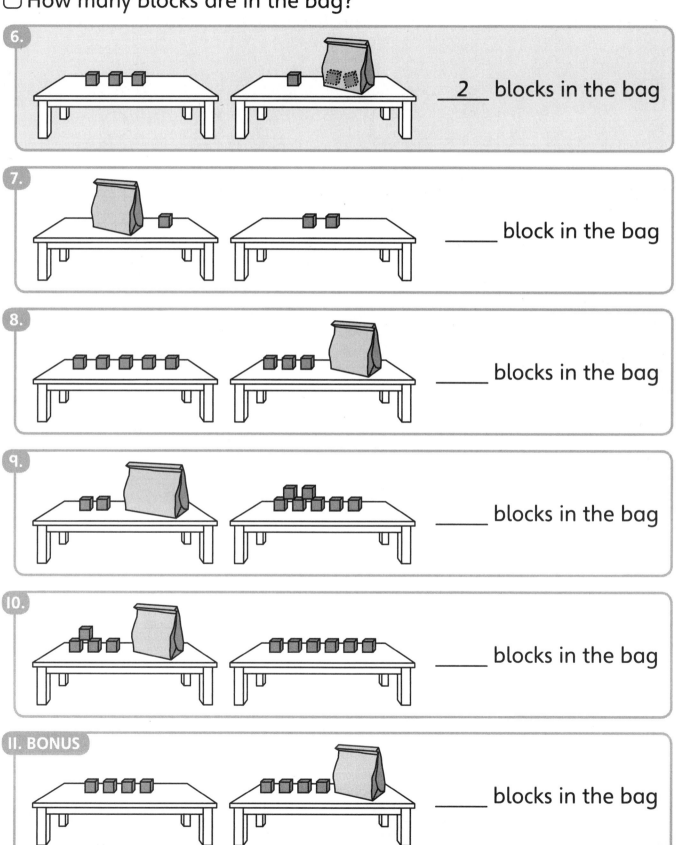

6. _2_ blocks in the bag

7. _____ block in the bag

8. _____ blocks in the bag

9. _____ blocks in the bag

10. _____ blocks in the bag

11. BONUS _____ blocks in the bag

The tables have the same number of balls.
☐ Draw the balls that are in the box.

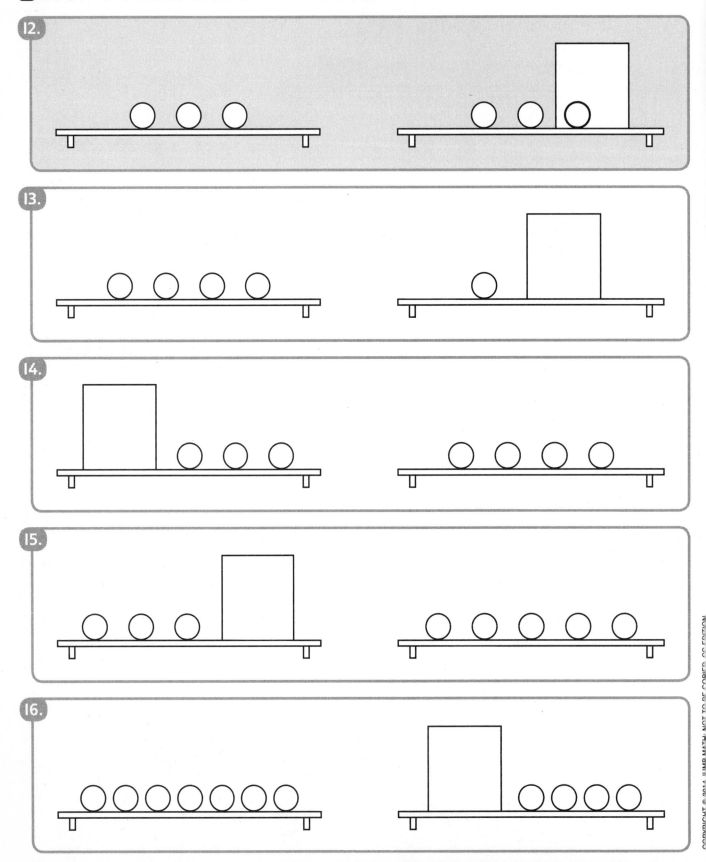

Operations and Algebraic Thinking 2-28

The tables have the same number of balls.

◯ Draw the balls that are in the box.

◯ Write the number of balls.

17.

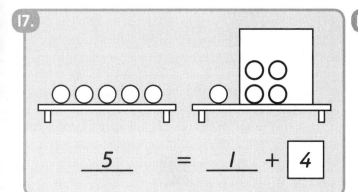

__5__ = __1__ + [4]

18.

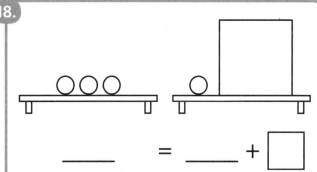

_____ = _____ + ☐

19.

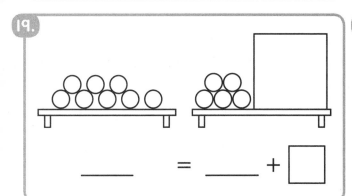

_____ = _____ + ☐

20.

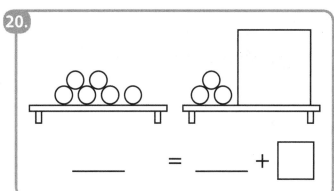

_____ = _____ + ☐

21.

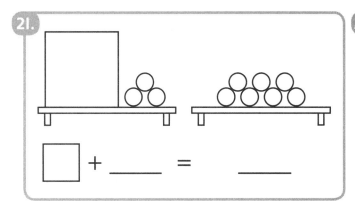

☐ + _____ = _____

22.

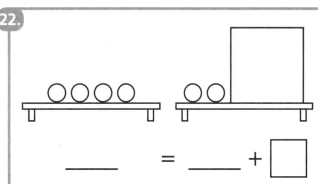

_____ = _____ + ☐

23.

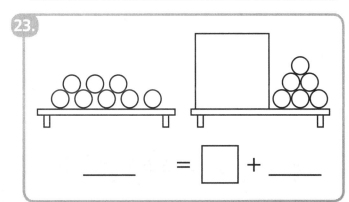

_____ = ☐ + _____

24.

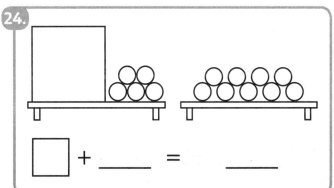

☐ + _____ = _____

☐ Circle the two numbers in the addition sentence.

☐ Draw dots for both numbers.

1.
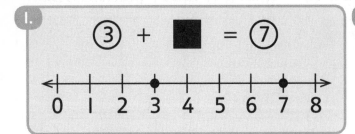

2.
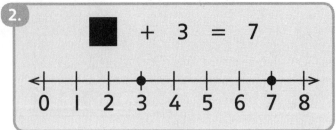

3.
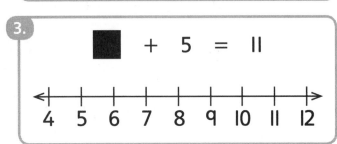

4.
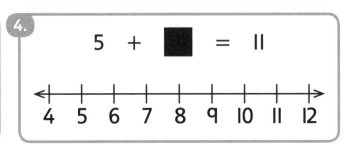

5.
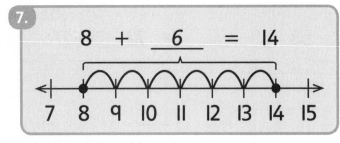

6.
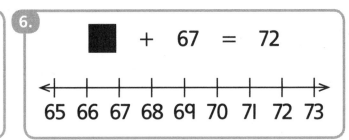

The missing addend is the number of jumps.

☐ Draw the dots.

☐ Find the missing addend by drawing the jumps.

7.
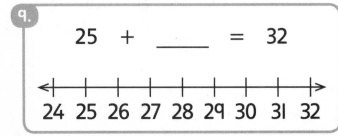

8.
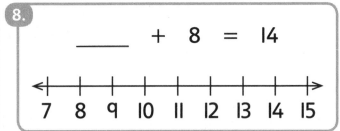

9.
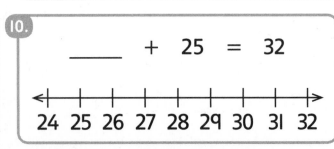

10.

___ + 25 = 32

Find the missing addend by using the number line.

11.

$\underline{\ \ 4\ \ } + 18 = 22$

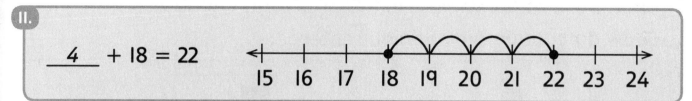

12.

$9 + \underline{\ \ \ \ \ } = 15$

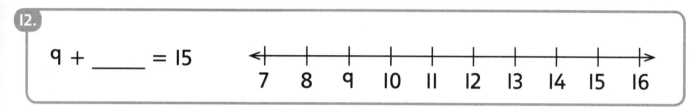

13.

$\underline{\ \ \ \ \ } + 26 = 33$

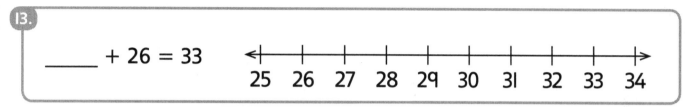

14.

$34 + \underline{\ \ \ \ \ } = 41$

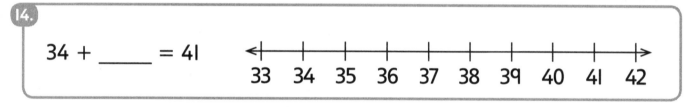

15.

$\underline{\ \ \ \ \ } + 45 = 52$

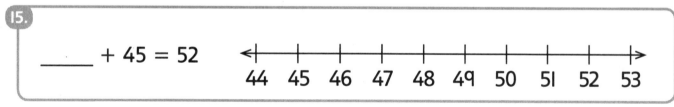

16.

$67 + \underline{\ \ \ \ \ } = 74$

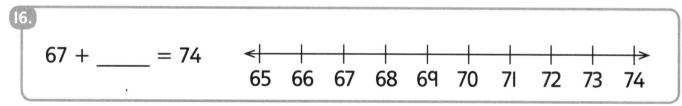

17.

$\underline{\ \ \ \ \ } + 81 = 88$

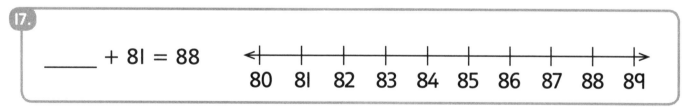

18.

$92 + \underline{\ \ \ \ \ } = 97$

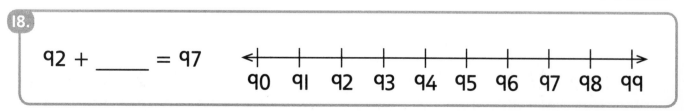

OA2-30 Find the Missing Addend by Counting On

◯ Write the addend you see in the ⬡.

◯ Circle the total in the addition sentence.

1.
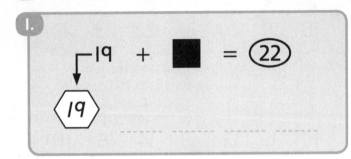

2.
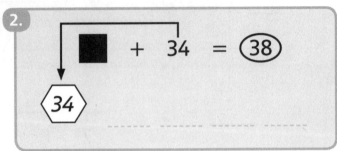

3.
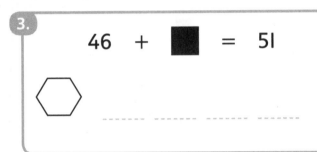

4.
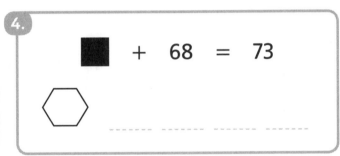

The missing addend is the number you count on.

◯ Find the missing addend by counting on to the total.

5.
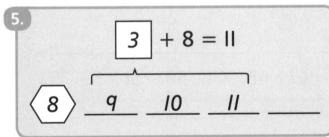

6.
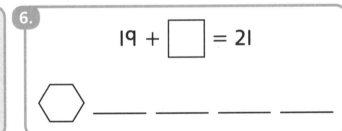

7.
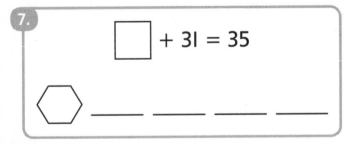

8.
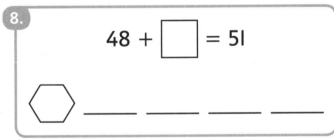

9.
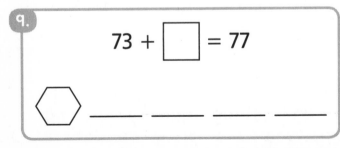

10.

○ Find the missing addend by counting on.

11.

$\boxed{5} + 7 = 12$

⟨7⟩ _8_ _9_ _10_ _11_ _12_ ___ ___ ___ ___ ___

12.

$16 + \boxed{} = 23$

⬡ ___ ___ ___ ___ ___ ___ ___ ___ ___ ___ ___

13.

$\boxed{} + 29 = 37$

⬡ ___ ___ ___ ___ ___ ___ ___ ___ ___ ___ ___

14.

$48 + \boxed{} = 57$

⬡ ___ ___ ___ ___ ___ ___ ___ ___ ___ ___ ___

15.

$\boxed{} + 67 = 72$

⬡ ___ ___ ___ ___ ___ ___ ___ ___ ___ ___ ___

16.

$85 + \boxed{} = 94$

⬡ ___ ___ ___ ___ ___ ___ ___ ___ ___ ___ ___

OA2-31 Word Problems with One Addend Missing (1)

⬭ Write the number sentence for the story.

1.

There are ■ red marbles.

There are 5 blue marbles.

There are 9 marbles altogether.

$$\begin{array}{r} + \ 5 \\ \hline 9 \end{array}$$

2.

There are 4 small dogs.

There are ■ large dogs.

There are 7 dogs altogether.

3.

There are 4 red cars.

There are ■ blue cars.

There are 6 cars altogether.

4.

There are ■ robins.

There are 5 owls.

There are 8 birds altogether.

5.

There are ■ picture books.

There are 2 story books.

There are 10 books altogether.

◯ Write the number sentence for the story.
◯ Write the missing addend.

6.

There are ⬚5 children at the park.

There are 3 adults at the park.

There are 8 people altogether.

$$\begin{array}{r} 5 \\ + 3 \\ \hline 8 \end{array}$$

7.

There are 7 glasses of milk.

There are ⬚ glasses of water.

There are 10 glasses altogether.

8.

Sarah has 4 stickers.

Ravi has ⬚ stickers.

Together, they have 10 stickers.

9.

Amy has ⬚ hockey cards.

Amy has 2 basketball cards.

Amy has 7 cards altogether.

10.

There are 2 open doors.

There are ⬚ closed doors.

There are 8 doors altogether.

OA2-32 Word Problems with One Addend Missing (2)

◯ Write the number sentence for the story.
◯ Write the missing addend.

1.

There were 3 rabbits on the grass.

[2] more rabbits came.

Now there are 5 rabbits on the grass.

$$\begin{array}{r} 3 \\ +\ \boxed{2} \\ \hline 5 \end{array}$$

2.

There were 14 toys in the box.

Sam put [   ] more toys in the box.

Now there are 17 toys in the box.

3.

Nina did 32 jumping jacks.

Then she did [   ] more jumping jacks.

Altogether, she did 36 jumping jacks.

4.

David counted 51 pens in the class.

Then he found [   ] more.

Altogether, he counted 56 pens.

5.

Tasha wrote 63 words.

Then she wrote [   ] more words.

Altogether, she wrote 68 words.

○ Write the number sentence for the story.
○ Write the missing addend.

6.

| 5 | birds flew to a tree. |

10 more birds flew to the tree.

Now there are 15 birds in the tree.

$$\begin{array}{r} 5 \\ +\ 10 \\ \hline 15 \end{array}$$

7.

☐ books were on the shelf.

Greg put 20 more books on the shelf.

Now there are 26 books on the shelf

8.

☐ names were in the jar.

May put 30 more names in the jar.

Altogether, there were 37 names in the jar.

9.

Ben read ☐ pages before school.

He read 32 more pages after school.

Altogether, he read 39 pages.

10.

☐ kids went to the zoo.

Then 41 more kids went to the zoo.

Now there are 47 kids at the zoo.

11.

The team won ☐ points.

Then the team won 71 more points.

Altogether, the team won 79 points.

OA2-33 Word Problems with Both Addends Missing

1.

Cathy has 2 balloons.

There are .

There are ◯.

How many of each are there?

$$\begin{array}{r} 0 \\ +\ 2 \\ \hline 2 \end{array}$$ 🎈 $$\begin{array}{r} 1 \\ +\ 1 \\ \hline 2 \end{array}$$ 🎈 $$\begin{array}{r} 2 \\ +\ 0 \\ \hline 2 \end{array}$$ 🎈

2.

Jin has 3 mugs.

There are 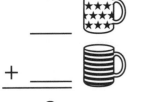. There are 🍶.

How many of each are there?

$$\begin{array}{r} \underline{\ \ \ } \\ +\ \underline{\ \ \ } \\ \hline 3 \end{array}$$ $$\begin{array}{r} \underline{\ \ \ } \\ +\ \underline{\ \ \ } \\ \hline 3 \end{array}$$ $$\begin{array}{r} \underline{\ \ \ } \\ +\ \underline{\ \ \ } \\ \hline 3 \end{array}$$ $$\begin{array}{r} \underline{\ \ \ } \\ +\ \underline{\ \ \ } \\ \hline 3 \end{array}$$

3.

Zara has 4 crayons.

There are ▮. There are ▮.

How many of each are there?

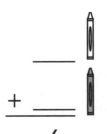

$$\begin{array}{r} \underline{\ \ \ } \\ +\ \underline{\ \ \ } \\ \hline 4 \end{array}$$ $$\begin{array}{r} \underline{\ \ \ } \\ +\ \underline{\ \ \ } \\ \hline 4 \end{array}$$ $$\begin{array}{r} \underline{\ \ \ } \\ +\ \underline{\ \ \ } \\ \hline 4 \end{array}$$ $$\begin{array}{r} \underline{\ \ \ } \\ +\ \underline{\ \ \ } \\ \hline 4 \end{array}$$ $$\begin{array}{r} \underline{\ \ \ } \\ +\ \underline{\ \ \ } \\ \hline 4 \end{array}$$

OA2-34 More Than and Counting On (1)

Find out which is more.

☐ Circle the extras.

1.
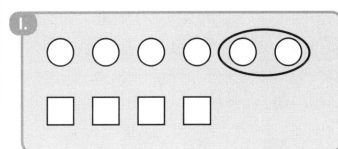

2.

3.
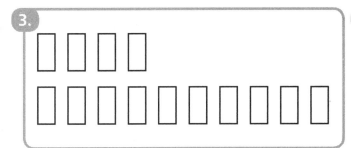

4.
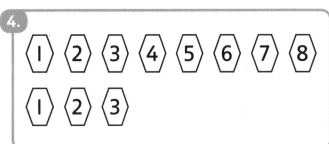

☐ Circle the extras.
☐ Write how many more.

5.

1 2 3 4 5 6 7 8 ⟨9 10⟩

1 2 3 4 5 6 7 8   ___2___ more

6.

1 2 3 4 5 6

1 2 3 4 5 6 7 8 9 10   _____ more

7.

1 2 3 4 5 6 7 8 9 10

1 2 3 4 5 6 7   _____ more

8.

1 2 3 4 5 6 7 8 9

1 2 3 4 5 6 7 8 9 10   _____ more

☐ Circle the extra numbers.
☐ Write how many more.

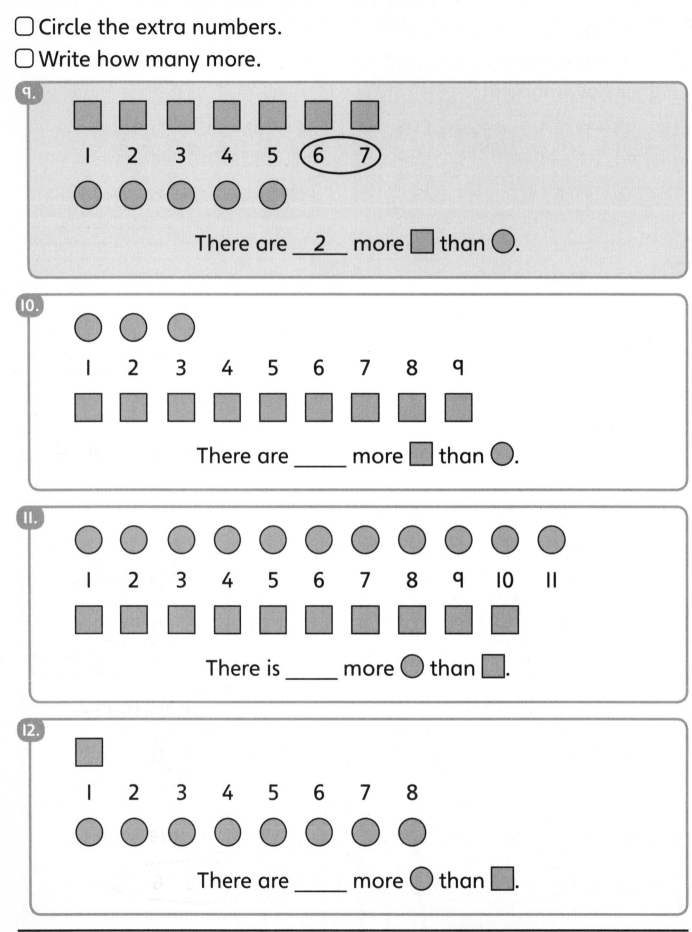

9.

1   2   3   4   5   ⑥   ⑦

There are __2__ more ⬜ than ◯.

10.

1   2   3   4   5   6   7   8   9

There are _____ more ⬜ than ◯.

11.

1   2   3   4   5   6   7   8   9   10   11

There is _____ more ◯ than ⬜.

12.

1   2   3   4   5   6   7   8

There are _____ more ◯ than ⬜.

Operations and Algebraic Thinking 2-34

◯ <u>Underline</u> the smaller number.
◯ Circle the numbers you count on.
◯ Write how many more.

13.
9 is ⬚3⬚ more than 6.

1 2 3 4 5 <u>6</u> (7 8 9)

14.
7 is ⬚ more than 5.

1 2 3 4 5 6 7

15.
5 is ⬚ more than 4.

1 2 3 4 5

16.
10 is ⬚ more than 6.

1 2 3 4 5 6 7 8 9 10

17.
8 is ⬚ more than 5.

1 2 3 4 5 6 7 8

18.
9 is ⬚ more than 3.

1 2 3 4 5 6 7 8 9

◯ Use the picture to complete the sentence.

19.
__6__ is ⬚2⬚ more than 4.

1 2 3 <u>4</u> (5 6)

20.
_____ is ⬚ more than 4.

1 2 3 <u>4</u> (5 6 7 8)

21.
_____ is ⬚ more than 7.

1 2 3 4 5 6 <u>7</u> (8 9)

22.
_____ is ⬚ more than 2.

1 <u>2</u> (3 4 5 6)

23.
_____ is ⬚ more than 3.

1 2 <u>3</u> (4 5 6 7 8)

24.
_____ is ⬚ more than 1.

<u>1</u> (2 3 4 5 6 7)

OA2-35 More Than and Counting On (2)

☐ Count on to find 4 more.

1.

5 __6__ __7__ __8__ __9__

__9__ is 4 more than 5.

2.

7 ____ ____ ____ ____

____ is 4 more than 7.

3.

4 ____ ____ ____ ____

____ is 4 more than 4.

4.

2 ____ ____ ____ ____

____ is 4 more than 2.

5.

6 ____ ____ ____ ____

____ is 4 more than 6.

6.

9 ____ ____ ____ ____

____ is 4 more than 9.

7.

8 ____ ____ ____ ____

____ is 4 more than 8.

8.

10 ____ ____ ____ ____

____ is 4 more than 10.

9.

3 ____ ____ ____ ____

____ is 4 more than 3.

10.

12 ____ ____ ____ ____

____ is 4 more than 12.

11.

15 ____ ____ ____ ____

____

12.

11 ____ ____ ____ ____

____

☐ Count on.

☐ Complete the sentence.

13.

7 more than 4 is __11__.

〈4〉 _5_ _6_ _7_ _8_ _9_ _10_ _11_ ___ ___ ___

14.

6 more than 9 is ____.

〈9〉 ___ ___ ___ ___ ___ ___ ___ ___ ___ ___

15.

8 more than 13 is ____.

〈13〉 ___ ___ ___ ___ ___ ___ ___ ___ ___ ___

16.

6 more than 27 is ____.

〈27〉 ___ ___ ___ ___ ___ ___ ___ ___ ___ ___

17.

5 more than 46 is ____.

〈46〉 ___ ___ ___ ___ ___ ___ ___ ___ ___ ___

18.

4 more than 59 is ____.

〈59〉 ___ ___ ___ ___ ___ ___ ___ ___ ___ ___

19.

5 more than 91 is ____.

〈91〉 ___ ___ ___ ___ ___ ___ ___ ___ ___ ___

OA2-36 Adding and How Many More

☐ Draw 3 more circles.
☐ Write the addition.

1.
3 more than 2
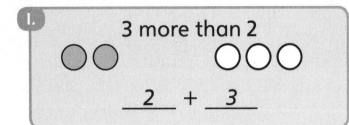
__2__ + __3__

2.
3 more than 1
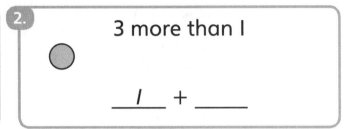
__1__ + _____

3.
3 more than 6
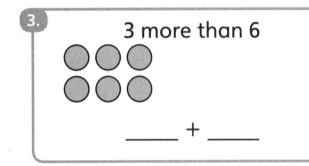
_____ + _____

4.
3 more than 3
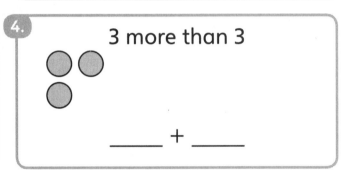
_____ + _____

☐ Draw more circles.
☐ Write the addition.

5.
1 more than 2
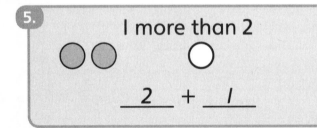
__2__ + __1__

6.
1 more than 4
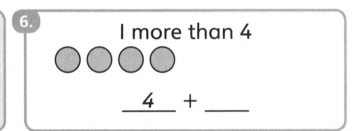
__4__ + _____

7.
4 more than 2
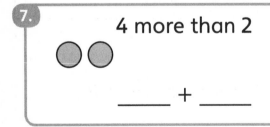
_____ + _____

8.
2 more than 3
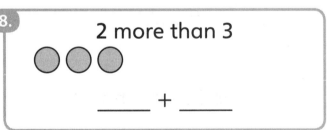
_____ + _____

9.
6 more than 1
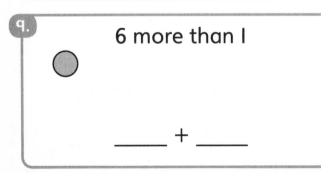
_____ + _____

10.
5 more than 4
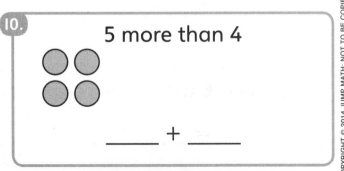
_____ + _____

○ Draw more dots to make 5.
○ Find the missing number.

11.

5 is __*1*__ more than 4.

5 = 4 + __*1*__

12.

5 is _____ more than 3.

5 = 3 + _____

13.

5 is _____ more than 2.

5 = 2 + _____

14.

5 is _____ more than 1.

5 = 1 + _____

○ Write the addition sentence.

15.

4 is 1 more than 3.

__*4*__ = __*3*__ + __*1*__

16.

8 is 3 more than 5.

_____ = __*5*__ + _____

17.

6 is 2 more than 4.

_____ = _____ + _____

18.

12 is 4 more than 8.

_____ = _____ + _____

19.

13 is 6 more than 7.

_____ = _____ + _____

20.

17 is 8 more than 9.

_____ = _____ + _____

OA2-37 Subtract Using Addition

☐ Write two subtraction sentences.
☐ Circle the totals.

1.
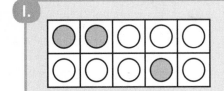

$\left(10\right) - \underline{\quad 3 \quad} = \underline{\quad 7 \quad}$

$\left(10\right) - \underline{\quad 7 \quad} = \underline{\quad 3 \quad}$

2.
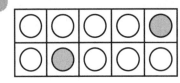

$10 - \underline{\quad 2 \quad} = \underline{\qquad}$

$10 - \underline{\quad 8 \quad} = \underline{\qquad}$

3.
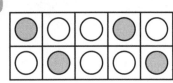

$10 - \underline{\qquad} = \underline{\qquad}$

$10 - \underline{\qquad} = \underline{\qquad}$

4.
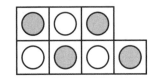

$\underline{\quad 7 \quad} - \underline{\qquad} = \underline{\qquad}$

$\underline{\quad 7 \quad} - \underline{\qquad} = \underline{\qquad}$

5.
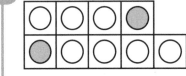

$\underline{\qquad} - \underline{\qquad} = \underline{\qquad}$

$\underline{\qquad} - \underline{\qquad} = \underline{\qquad}$

6.

$\underline{\qquad} - \underline{\qquad} = \underline{\qquad}$

$\underline{\qquad} - \underline{\qquad} = \underline{\qquad}$

☐ Write two addition sentences.
☐ Write two subtraction sentences.
☐ Circle the totals.

7.
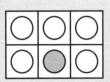

$\underline{\quad 1 \quad} + \underline{\quad 5 \quad} = \left(6\right)$

$\underline{\quad 5 \quad} + \underline{\quad 1 \quad} = \left(6\right)$

$\left(6\right) - \underline{\quad 5 \quad} = \underline{\quad 1 \quad}$

$\left(6\right) - \underline{\quad 1 \quad} = \underline{\quad 5 \quad}$

8.
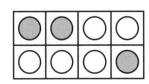

$\underline{\qquad} + \underline{\qquad} = \underline{\qquad}$

$\underline{\qquad} + \underline{\qquad} = \underline{\qquad}$

$\underline{\qquad} - \underline{\qquad} = \underline{\qquad}$

$\underline{\qquad} - \underline{\qquad} = \underline{\qquad}$

9.
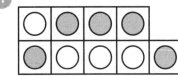

$\underline{\qquad} + \underline{\qquad} = \underline{\qquad}$

$\underline{\qquad} + \underline{\qquad} = \underline{\qquad}$

$\underline{\qquad} - \underline{\qquad} = \underline{\qquad}$

$\underline{\qquad} - \underline{\qquad} = \underline{\qquad}$

Operations and Algebraic Thinking 2-37

○ Circle the total.
○ Write two subtraction sentences for the addition.

10.
$$3 + 7 = \boxed{10}$$
$$10 - 3 = 7$$
$$10 - 7 = 3$$

11.
$$2 + 5 = \boxed{7}$$
$$7 - 2 = 5$$
_____

12.
$$3 + 6 = \boxed{9}$$
_____
_____

13.
$$9 + 2 = 11$$
_____
_____

14.
$$12 = 5 + 7$$
_____
_____

15.
$$13 = 4 + 9$$
_____
_____

○ Circle the total.
○ Write two addition sentences for the subtraction.

16.
$$\boxed{8} - 3 = 5$$
$$3 + 5 = 8$$
$$5 + 3 = 8$$

17.
$$\boxed{12} - 4 = 8$$
$$4 + 8 = 12$$
_____

18.
$$\boxed{7} - 3 = 4$$
_____
_____

19.
$$11 - 6 = 5$$
_____
_____

20.
$$7 = 9 - 2$$
_____
_____

21.
$$8 = 15 - 7$$
_____
_____

$$1 + 6 = 7 \qquad 1 + 7 = 8 \qquad 1 + 8 = 9$$
$$2 + 5 = 7 \qquad 2 + 6 = 8 \qquad 2 + 7 = 9$$
$$3 + 4 = 7 \qquad 3 + 5 = 8 \qquad 3 + 6 = 9$$
$$\qquad\qquad\quad 4 + 4 = 8 \qquad 4 + 5 = 9$$

☐ Which fact do you use?

22.
$$7 - 2$$
Use $2 + \underline{5} = 7$

23.
$$8 - 1$$
Use $1 + \underline{} = 8$

24.
$$9 - 3$$
Use $3 + \underline{} = 9$

25.
$$9 - 5$$
Use $\underline{4} + 5 = 9$

26.
$$7 - 6$$
Use $\underline{} + 6 = 7$

27.
$$8 - 5$$
Use $\underline{} + 5 = 8$

☐ Use an addition fact to subtract.

28.
$$7 - 2$$
$2 + \underline{5} = 7$
So $7 - 2 = \underline{5}$

29.
$$8 - 6$$
$6 + \underline{2} = 8$
So $8 - 6 = \underline{}$

30.
$$7 - 4$$
$4 + \underline{} = 7$
So $7 - 4 = \underline{}$

31.
$$9 - 2$$
$2 + \underline{} = 9$
So $9 - 2 = \underline{}$

32.
$$8 - 4$$
$4 + \underline{} = 8$
So $8 - 4 = \underline{}$

33.
$$10 - 2$$
$2 + \underline{} = 10$
So $10 - 2 = \underline{}$

34.
$$10 - 4 \qquad 8 - 3 \qquad 9 - 4 \qquad 6 - 1 \qquad 12 - 8 \qquad 11 - 9$$

Operations and Algebraic Thinking 2-37

OA2-38 Subtract by Counting On

Will wants to find 6 − 2 = ___. He counts on to solve 2 + ___ = 6.

2   3   4   5   6

The answer is the number of fingers. So 6 − 2 = _4_ .

⬭ Find the missing number by counting on.

1.
7 − 5 = _2_

5 + _2_ = 7

2.
5 − 4 = ___

4 + ___ = 5

3.
5 − 3 = ___

3 + ___ = 5

4.
9 − 2 = ___

2 + ___ = 9

5.
11 − 8 = _3_

8 + _3_ = 11

6.
12 − 7 = ___

7 + ___ = 12

7.
15 − 13 = ___

13 + ___ = 15

8.
27 − 22 = ___

22 + ___ = 27

9.
12 − 9 = ___

9 + ___ = 12

10.
7 − 4 = ___

4 + ___ = 7

11.
8 − 6 = ___

6 + ___ = 8

12.
15 − 11 = ___

11 + ___ = 15

⬭ Subtract.

13.
9 − 7 = ___

14.
5 − 2 = ___

15.
12 − 8 = ___

16.
11 − 4 = ___

17.
6 − 5 = ___

18.
8 − 5 = ___

19.
14 − 12 = ___

20.
26 − 23 = ___

21.
17 − 11 = ___

22.
35 − 33 = ___

23.
19 − 16 = ___

24.
15 − 9 = ___

Operations and Algebraic Thinking 2-38

OA2-39 Subtracting and How Many More

☐ Shade how many to take away. Then subtract.

1.
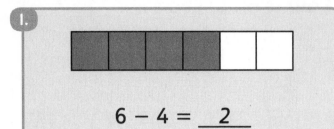

6 − 4 = __2__

2.
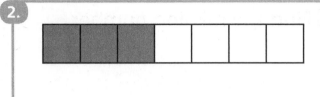

7 − 3 = _____

3.
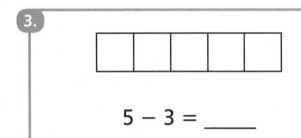

5 − 3 = _____

4.
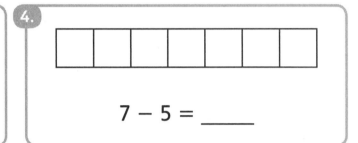

7 − 5 = _____

☐ Count how many more.
☐ Find the missing number.

5.
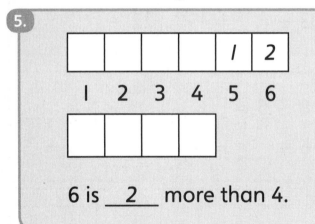

6 is __2__ more than 4.

6.
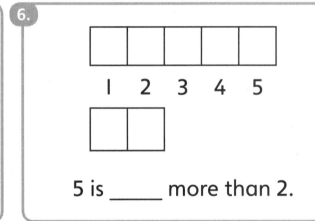

5 is _____ more than 2.

7.
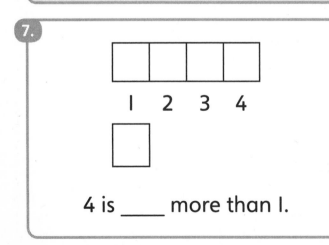

4 is _____ more than 1.

8.
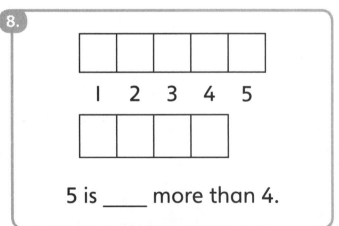

5 is _____ more than 4.

◯ Shade how many to take away.
◯ Count how many more.
◯ Find the missing number.

9.

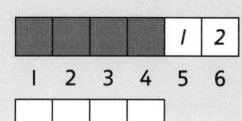

1   2   3   4   5   6

6 is __2__ more than 4.

6 − 4 = __2__

10.

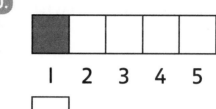

1   2   3   4   5

5 is ____ more than 1.

5 − 1 = ____

11.

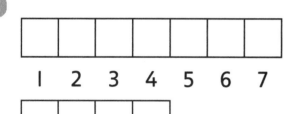

1   2   3   4   5   6   7

7 is ____ more than 4.

7 − 4 = ____

12.

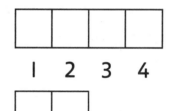

1   2   3   4

4 is ____ more than 2.

4 − 2 = ____

13.

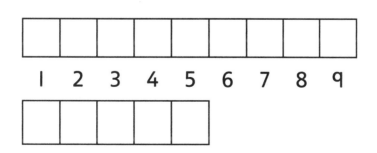

1   2   3   4   5   6   7   8   9

9 is ____ more than 5.

9 − 5 = ____

Operations and Algebraic Thinking 2-39

OA2-40 Compare Using Subtraction

◯ Use the word sentence to subtract.

1.
6 is **2** more than 4.

6 − 4 = ___2___

2.
5 is **3** more than 2.

5 − 2 = _____

3.
15 is **7** more than 8.

15 − 8 = _____

4.
95 is **12** more than 83.

95 − 83 = _____

◯ Subtract.

5.
7 is how many more than 5?

___7___ − ___5___ = _____

6.
8 is how many more than 3?

___8___ − _____ = _____

7.
13 is how many more than 9?

_____ − _____ = _____

8.
5 is how many more than 3?

_____ − _____ = _____

9.
17 is how many more than 14?

_____ − _____ = _____

10.
16 is how many more than 2?

_____ − _____ = _____

11.
9 is how many more than 6?

_____ − _____ = _____

12.
13 is how many more than 3?

_____ − _____ = _____

13. BONUS
25 is how many more than 21?

14. BONUS
36 is how many more than 29?

☐ Write the subtraction.

15.

Sara has 7 apples. Anna has 5 apples.

Sara has ___7 − 5___ more apples.

16.

Marco has 6 crayons. Amit has 4 crayons.

Marco has ___6 − ___ more crayons.

17.

Rani has 12 marbles. Jayden has 10 marbles.

Rani has _____ more marbles.

☐ Write the subtraction.
☐ Subtract.

18.

Kate has 7 stamps. Ben has 5 stamps.

How many more stamps does Kate have?

___7 − 5___ = ___2___

19.

Lily has 9 stickers. Peter has 6 stickers.

How many more stickers does Lily have?

___9 − ___ = _____

20.

Tom has 17 balloons. Zara has 7 balloons.

How many more balloons does Tom have?

_____ = _____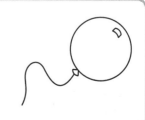

OA2-41 Missing Numbers in Subtraction

○ Write the total number of boxes.
○ Write two subtraction sentences.

1.

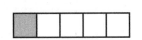

Total __7__

__7__ – __2__ = __5__

__7__ – __5__ = __2__

2.

Total __5__

__5__ – __1__ = __4__

_____ – _____ = _____

3.

Total _____

_____ – _____ = _____

_____ – _____ = _____

4.

Total _____

_____ – _____ = _____

_____ – _____ = _____

To find 8 – ☐ = 5, use 8 – __5__ = ☐3. So 8 – ☐3 = 5.

○ Find the missing numbers.

5.

8 – ☐3 = 5

8 – __5__ = ☐3

6.

6 – ☐ = 4

6 – _____ = ☐

7.

7 – ☐ = 4

7 – _____ = ☐

8.

12 – ☐ = 8

12 – _____ = ☐

9.

9 – ☐ = 3

9 – _____ = ☐

10.

10 – ☐ = 6

10 – _____ = ☐

Operations and Algebraic Thinking 2-41

☐ Color the answer.

11.
8 bunnies were eating grass. Some hopped away.
6 bunnies kept eating. How many hopped away?

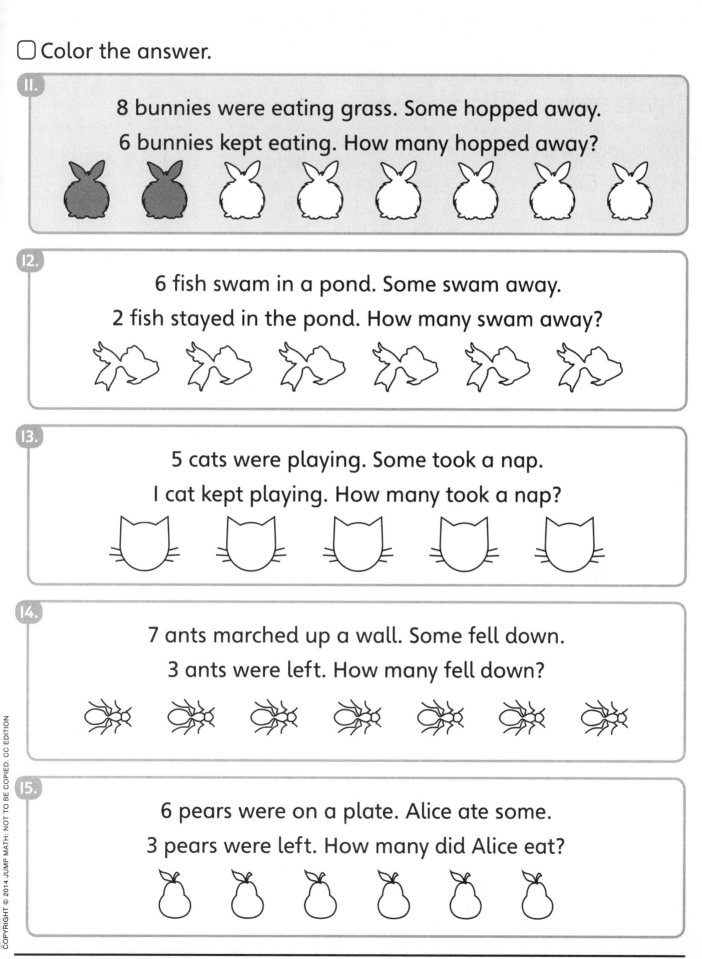

12.
6 fish swam in a pond. Some swam away.
2 fish stayed in the pond. How many swam away?

13.
5 cats were playing. Some took a nap.
I cat kept playing. How many took a nap?

14.
7 ants marched up a wall. Some fell down.
3 ants were left. How many fell down?

15.
6 pears were on a plate. Alice ate some.
3 pears were left. How many did Alice eat?

OA2-42 Missing Total in Subtraction

☐ Draw the dots on the big domino.

☐ Write the total.

1.

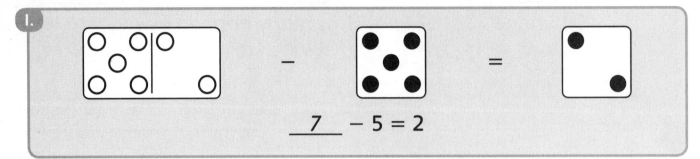

$\underline{7} - 5 = 2$

2.

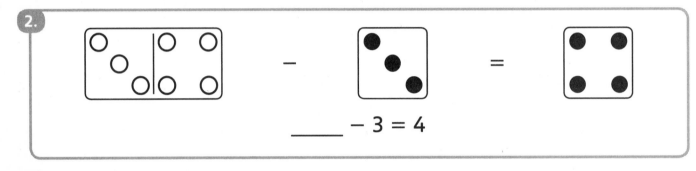

$\underline{} - 3 = 4$

3.

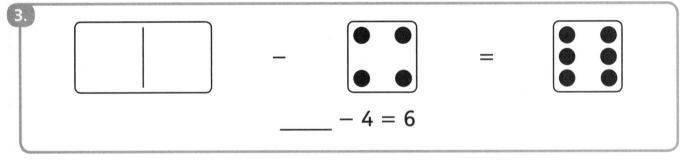

$\underline{} - 4 = 6$

4.

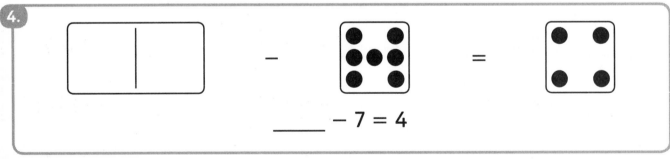

$\underline{} - 7 = 4$

5.

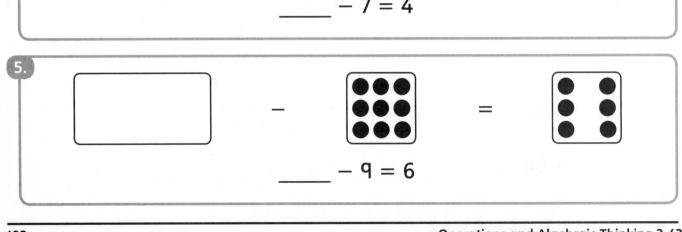

$\underline{} - 9 = 6$

◯ Find the missing total.

6.
$12 + 4 = \underline{16}$

so

$\underline{16} - 4 = 12$

7.
$14 + 3 = \underline{}$

so

$\underline{} - 3 = 14$

8.
$5 + 5 = \underline{}$

so

$\underline{} - 5 = 5$

9.
$12 + 5 = \underline{}$

so

$\underline{} - 5 = 12$

10.
$3 + 14 = \underline{}$

so

$\underline{} - 14 = 3$

11.
$7 + 6 = \underline{}$

so

$\underline{} - 6 = 7$

12.
$27 + 4 = \underline{}$

so

$\underline{} - 4 = 27$

13.
$36 + 3 = \underline{}$

so

$\underline{} - 3 = 36$

14.
$2 + 25 = \underline{}$

so

$\underline{} - 25 = 2$

◯ Find the missing total.

15.

$\underline{} - 3 = 5$ $\underline{} - 2 = 8$ $\underline{} - 2 = 7$

$\underline{} - 4 = 15$ $\underline{} - 5 = 6$ $\underline{} - 22 = 5$

$\underline{} - 2 = 32$ $\underline{} - 26 = 3$ $\underline{} - 6 = 7$

$\underline{} - 45 = 2$ $\underline{} - 2 = 56$ $\underline{} - 78 = 1$

$\underline{} - 64 = 3$ $\underline{} - 5 = 39$ $\underline{} - 82 = 2$

OA2-43 Subtraction Word Problems

○ Write a number sentence for the story.
○ Find the missing number.

1.

There were $\boxed{8}$ apples in the tree.

3 apples fell down.

Now there are 5 apples in the tree.

$$\begin{array}{r} \boxed{8} \\ -3 \\ \hline 5 \end{array}$$

2.

There were $\boxed{}$ rabbits on the grass.

5 rabbits hopped away.

Now there are 2 rabbits on the grass.

3.

There were $\boxed{}$ books on the shelf.

Miss Chen took 4 books.

Now there are 7 books on the shelf.

4.

Nina had 10 raisins.

She ate $\boxed{}$ raisins.

Nina has 6 raisins left.

$$\begin{array}{r} 10 \\ -\boxed{} \\ \hline 6 \end{array}$$

5.

There were 9 toys in the toy box.

Roy took out $\boxed{}$ toys.

Now there are 6 toys in the toy box.

○ Write the number sentence for the story.
○ Write the answer.

6.

Ravi had 8 berries.

He ate some berries.

Now Ravi has 3 berries left.

How many berries did he eat? ___5___

$$\begin{array}{r} 8 \\ -5 \\ \hline 3 \end{array}$$

7.

Kim had 9 books.

She gave some away.

Now she has 7 books.

How many books did Kim give away? _____

8.

There were 8 apples in the tree.

Some apples fell.

Now there are 2 apples in the tree.

How many apples fell? _____

9.

There were some birds in a tree.

Then 5 birds flew away.

Now there are 2 birds in the tree.

How many birds were in the tree before? _____

10.

There were some frogs in the pond.

Then 3 frogs hopped out.

Now there are 6 frogs in the pond.

How many frogs were in the pond before? _____

NBT2-1 Tens and Ones Blocks

☐ What number do the ones show?

1.

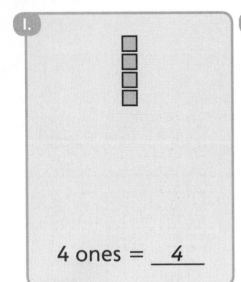

4 ones = ___4___

2.

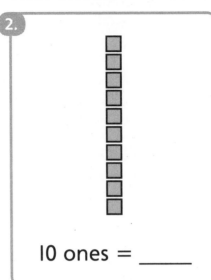

10 ones = _____

3.

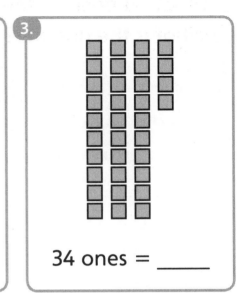

34 ones = _____

☐ Write the number of tens.

4.

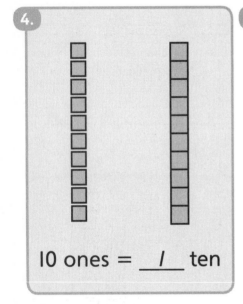

10 ones = ___1___ ten

5.

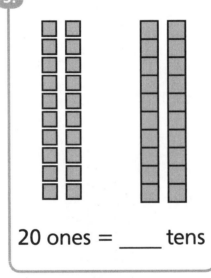

20 ones = _____ tens

6.

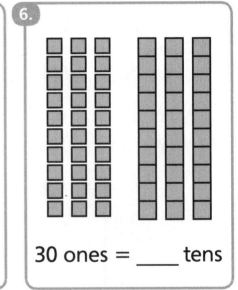

30 ones = _____ tens

☐ Write the number.

7. 4 tens = ___40___

8. 6 tens = _____

9. 8 tens = _____

10. 9 tens = _____

11. 7 tens = _____

12. 5 tens = _____

☐ What number do the tens show?
☐ What number do the ones show?
☐ What is the number?

13.

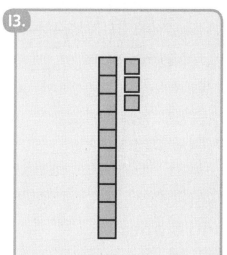

I ten = _10_

3 ones = _3_

Number = _13_

14.

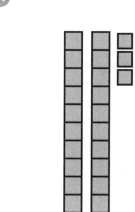

2 tens = _____

3 ones = _____

Number = _____

15.

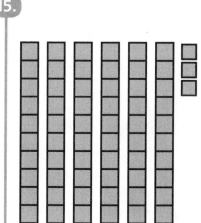

6 tens = _____

3 ones = _____

Number = _____

16.

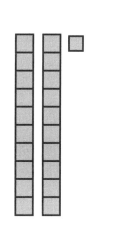

2 tens = _____

1 one = _____

Number = _____

17.

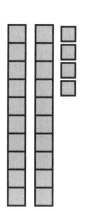

2 tens = _____

4 ones = _____

Number = _____

18.

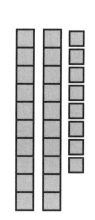

2 tens = _____

8 ones = _____

Number = _____

Number and Operations in Base Ten 2-1

◻ Use the number of tens and ones to write the number.

19.
3 tens = __30__

8 ones = __8__

Number = __38__

20.
5 tens = _____

7 ones = _____

Number = _____

21.
9 tens = _____

2 ones = _____

Number = _____

22.
6 tens and 1 one = __61__

23.
5 tens and 4 ones = _____

24.
2 tens and 2 ones = _____

25.
7 tens and 3 ones = _____

◻ Circle ones and tens to show the number.

26.
53
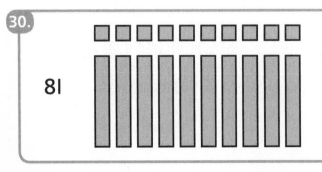

27.
15
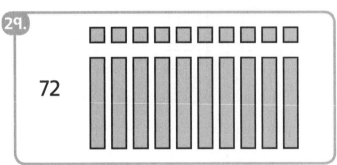

28.
68
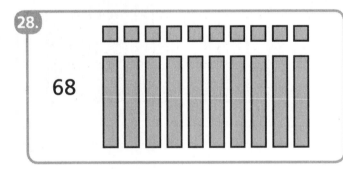

29.
72

30.
81
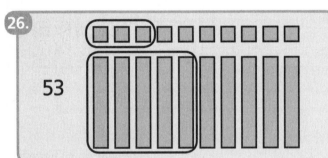

31.
97
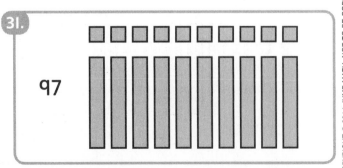

Number and Operations in Base Ten 2-1

☐ Draw tens (|) and ones (•) to show the number.

32.
23
|| :
    •

33.
17

34.
32

35.
46

36.
51

37.
75

☐ Circle the picture that shows the number.

38.
12

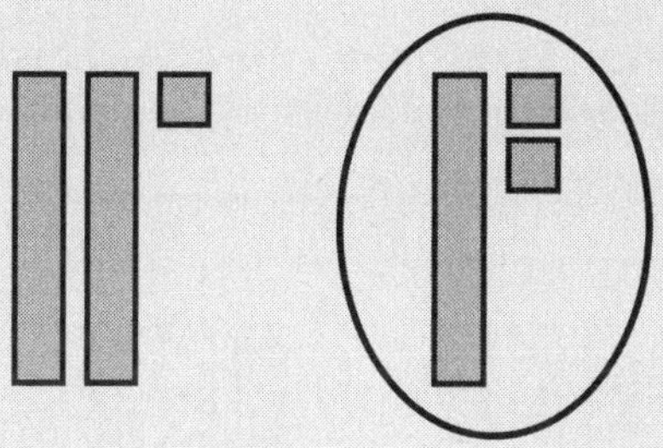

39.
34

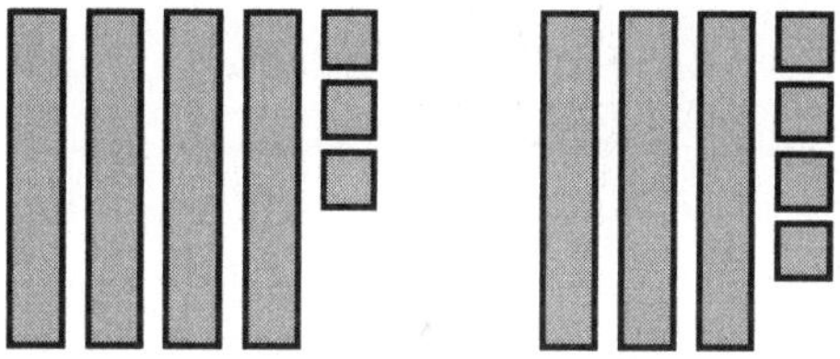

40.
23

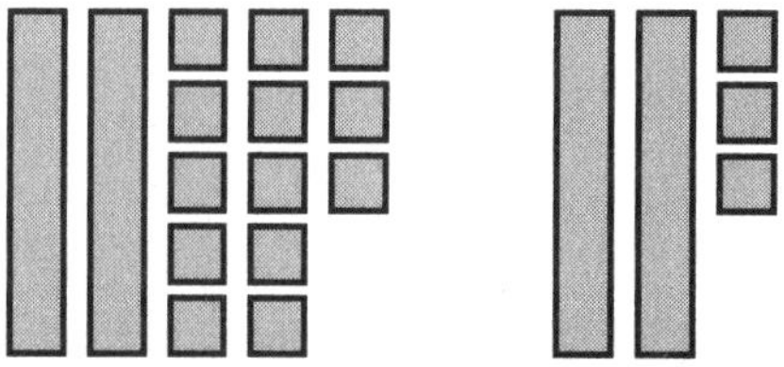

41.
41

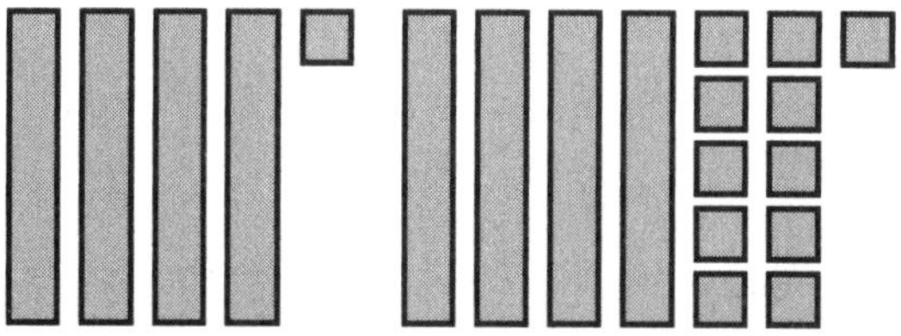

42. BONUS
87

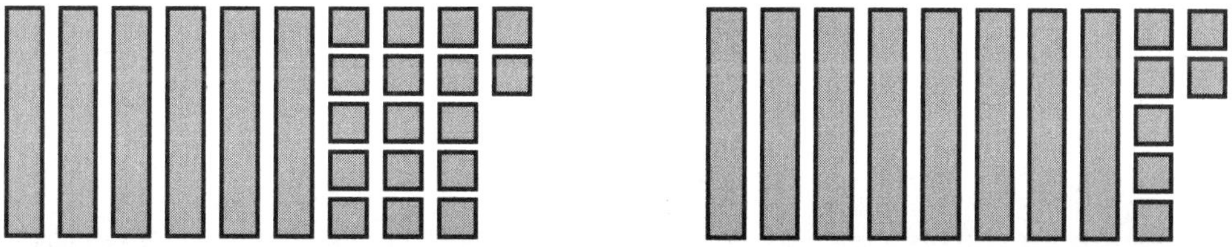

Circle the picture that does **not** show the number.

43.

42

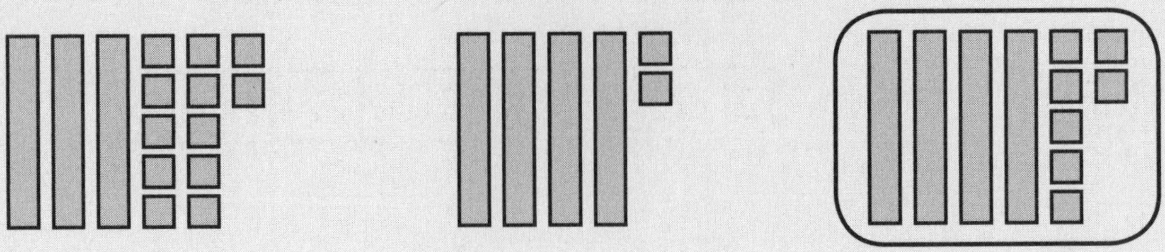

44.

56

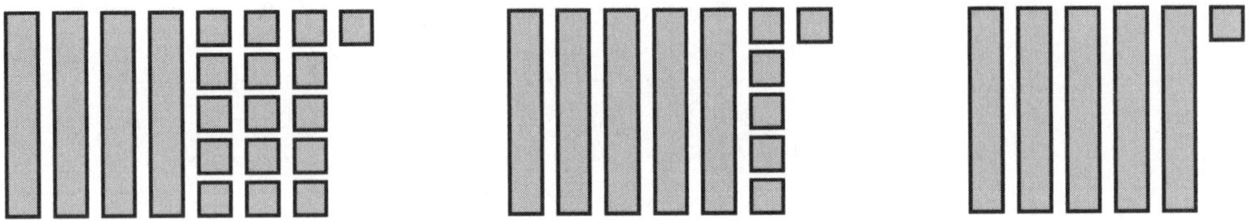

45.

33

46.

67

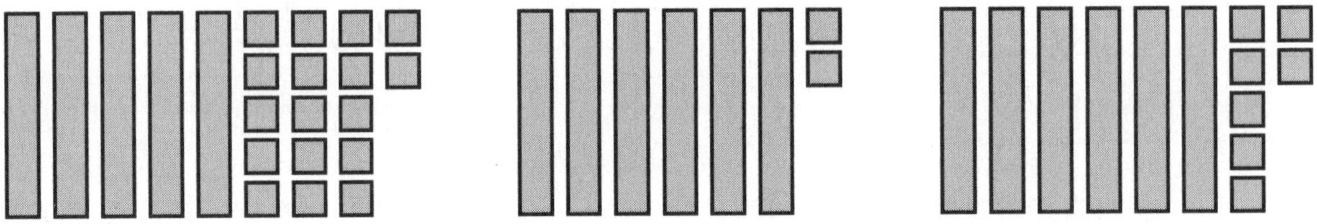

47.

28

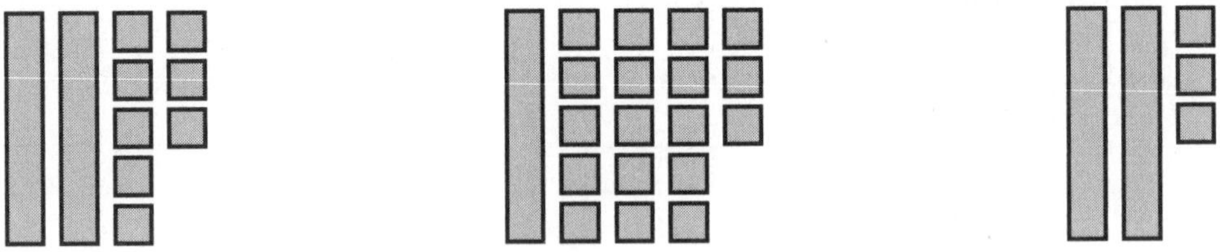

48.

71

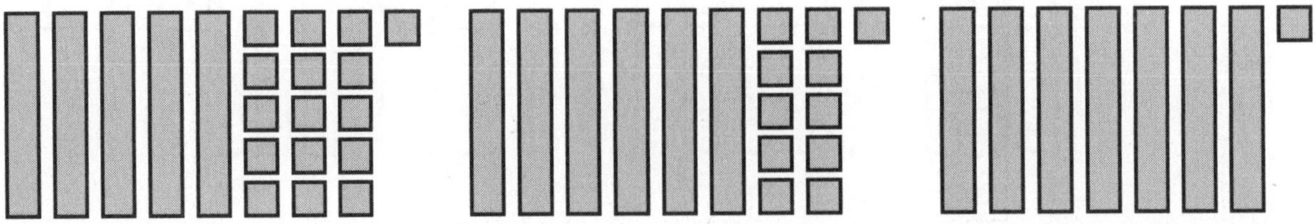

Number and Operations in Base Ten 2-1

NBT2-2 Hundreds Charts

☐ Shade the numbers that are **smaller** or **larger**.

1. larger

| 21 | 22 | 23 | 24 | 25 | 26 | **27** | 28 | 29 | 30 |

2. smaller

| 11 | 12 | 13 | 14 | 15 | 16 | 17 | 18 | **19** | 20 |

3. smaller

| 81 | 82 | 83 | 84 | 85 | **86** | 87 | 88 | 89 | 90 |

4. larger

| 31 | 32 | **33** | 34 | 35 | 36 | 37 | 38 | 39 | 40 |

5. smaller

| 51 | **52** | 53 | 54 | 55 | 56 | 57 | 58 | 59 | 60 |

6. larger

| 41 | 42 | 43 | 44 | 45 | 46 | 47 | 48 | **49** | 50 |

☐ Shade the numbers that are **smaller** or **larger**.

7. larger	8. smaller	9. smaller	10. larger	11. larger	12. smaller
1	1	3	3	8	8
11	11	13	13	18	**18**
21	21	**23**	**23**	28	28
31	31	33	33	38	38
41	41	43	43	48	48
51	51	53	53	58	58
61	**61**	63	63	68	68
71	71	73	73	**78**	78
81	81	83	83	88	88

◻ Shade the numbers that are smaller or larger.

13.

larger

11	12	13	14	15	16	17	18	19	**20**
21	22	23	24	25	26	27	28	29	30

14.

smaller

11	12	13	14	15	16	17	18	19	**20**
21	22	23	24	25	26	27	28	29	30

15.

smaller

41	42	43	44	45	46	47	48	49	50
51	52	53	54	55	**56**	57	58	59	60

16.

larger

41	42	43	44	45	46	47	48	**49**	50
51	52	53	54	55	56	57	58	59	60

◻ Use the hundreds chart to fill in the missing numbers.

61	62	63	64	65	66	67	68	69	70
71	72	73	74	75	76	77	78	79	80
81	82	83	84	85	86	87	88	89	90
91	92	93	94	95	96	97	98	99	100

17.

63	64	65

18.

	74	

19.

	84	

20.

	70	

21.

	80	

22.

	90	

Tom circles all the numbers that have a 4.

He shades all the numbers that have a 9.

1	2	3	(4)	5	6	7	8	**9**	10
11	12	13	(14)	15	16	17	18	**19**	20
21	22	23	(24)	25	26	27	28	**29**	30

☐ Circle all the numbers that have a 1.

☐ Shade all the numbers that have an 8.

23.

31	32	33	34	35	36	37	38	39	40
41	42	43	44	45	46	47	48	49	50
51	52	53	54	55	56	57	58	59	60

☐ Circle all the numbers that have a 5.

☐ Shade all the numbers that have a 0.

24.

71	72	73	74	75	76	77	78	79	80
81	82	83	84	85	86	87	88	89	90
91	92	93	94	95	96	97	98	99	100

☐ **BONUS:** Shade all the numbers that have a 7.

25.

61	62	63	64	65	66	67	68	69	70
71	72	73	74	75	76	77	78	79	80
81	82	83	84	85	86	87	88	89	90

☐ In a hundreds chart, what number is above?
☐ What number is below?

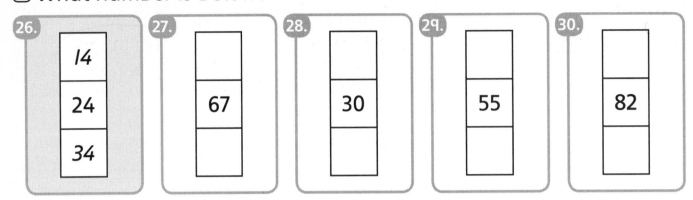

26.
14
24
34

27.
67

28.
30

29.
55

30.
82

☐ In a hundreds chart, what number is before?
☐ What number is after?

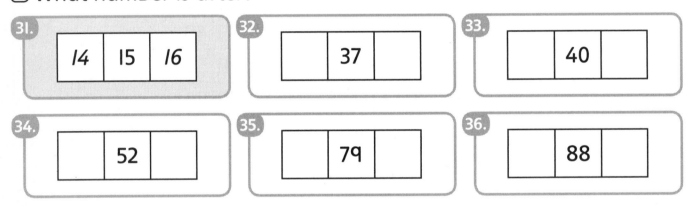

31.
| 14 | 15 | 16 |

32.
| | 37 | |

33.
| | 40 | |

34.
| | 52 | |

35.
| | 79 | |

36.
| | 88 | |

☐ Fill in the missing numbers from the hundreds chart.

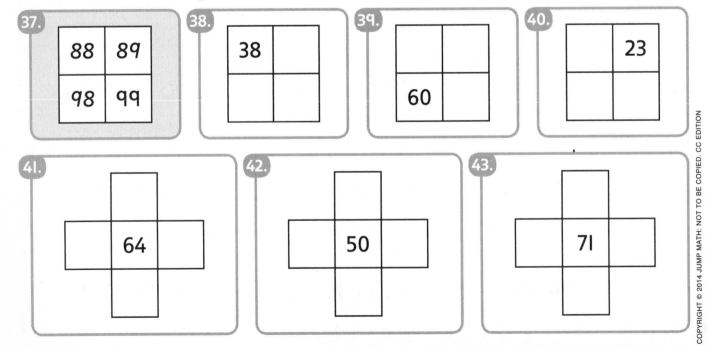

37.
88	89
98	99

38.
38	

39.
60	

40.
	23

41.
64

42.
50

43.
71

NBT2-3 Ordering 2-Digit Numbers

☐ Circle the group that is in counting order.

1.
18, 20, 19   20, 18, 19   ⟨18, 19, 20⟩

2.
35, 37, 36   35, 36, 37   37, 36, 35

3.
88, 89, 90   90, 88, 89   88, 90, 89

4.
50, 51, 49   49, 50, 51   51, 49, 50

☐ Circle the group that is in counting order.

5.
11, 13, 12, 14   ⟨11, 12, 13, 14⟩   13, 11, 14, 12

6.
29, 27, 30, 28   28, 29, 30, 27   27, 28, 29, 30

7.
89, 90, 91, 92   90, 91, 89, 92   89, 91, 90, 92

8.
77, 76, 74, 75   74, 75, 76, 77   77, 74, 75, 76

☐ Circle the numbers on the number line.
☐ Write the numbers in order from **smallest** to **largest**.

9.

23, 12, 17

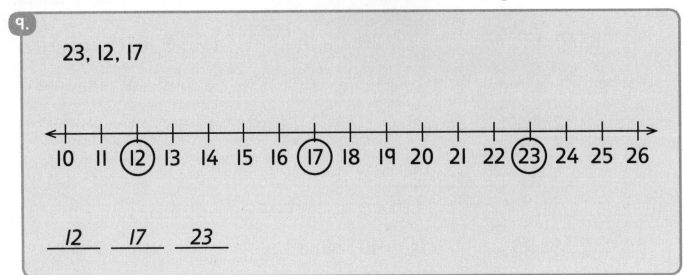

__12__   __17__   __23__

10.

49, 60, 51

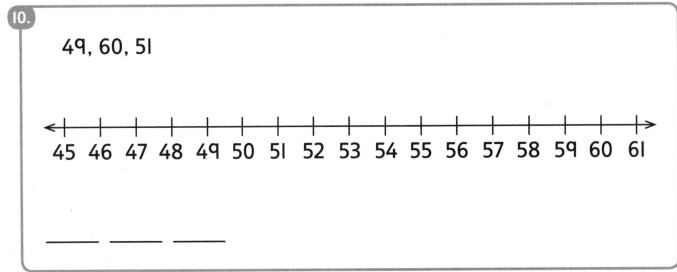

_____ _____ _____

11.

81, 92, 78

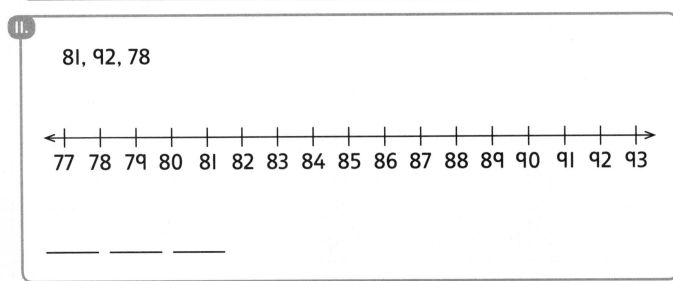

_____ _____ _____

Number and Operations in Base Ten 2-3

NBT2-4 Expanded Form

☐ Write the number.

1. 9 tens and 4 ones = __94__

2. 2 tens and 7 ones = _____

3. 5 tens and 3 ones = _____

4. 8 tens and 8 ones = _____

5. 1 ten and 5 ones = _____

6. 7 tens and 2 ones = _____

7. 4 tens and 6 ones = _____

8. 6 tens and 1 one = _____

9. 3 tens and 9 ones = _____

10. 9 tens and 0 ones = _____

☐ Write the number in **expanded form**.

11.

1	2	3	4	5	6	7	8	9	10
11	12	13	14	15	16	17	18	19	20
21	22	23	24	25	26	27	28	29	30

__24__ = __20__ + __4__

12.

1	2	3	4	5	6	7	8	9	10
11	12	13	14	15	16	17	18	19	20
21	22	23	24	25	26	27	28	29	30
31	32	33	34	35	36	37	38	39	40

_____ = _____ + _____

☐ Fill in the blanks for the expanded form.
☐ Fill in the blanks for the base ten names.

13.

12 = __10__ + __2__   12 = __1__ ten and __2__ ones

12 = __2__ + __10__   12 = __2__ ones and __1__ ten

14.

17 = ____ + ____   17 = ____ ten and ____ ones

17 = ____ + ____   17 = ____ ones and ____ ten

15.

35 = ____ + ____   35 = ____ tens and ____ ones

35 = ____ + ____   35 = ____ ones and ____ tens

16.

48 = ____ + ____   48 = ____ tens and ____ ones

48 = ____ + ____   48 = ____ ones and ____ tens

17.

26 = ____ + ____   26 = ____ tens and ____ ones

26 = ____ + ____   26 = ____ ones and ____ tens

18.

97 = ____ + ____   97 = ____ tens and ____ ones

97 = ____ + ____   97 = ____ ones and ____ tens

☐ Count on to add.

1.

$45 + 3 = \underline{\ ?\ }$ 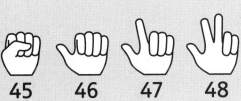 $45 + 3 = \underline{\ 48\ }$

45   46   47   48

2.

$16 + 2 = \underline{\hspace{1cm}}$ $16 + 4 = \underline{\hspace{1cm}}$ $16 + 7 = \underline{\hspace{1cm}}$

3.

$26 + 2 = \underline{\hspace{1cm}}$ $36 + 4 = \underline{\hspace{1cm}}$ $46 + 7 = \underline{\hspace{1cm}}$

4.

$56 + 2 = \underline{\hspace{1cm}}$ $56 + 4 = \underline{\hspace{1cm}}$ $56 + 7 = \underline{\hspace{1cm}}$

5.

$86 + 2 = \underline{\hspace{1cm}}$ $86 + 4 = \underline{\hspace{1cm}}$ $86 + 7 = \underline{\hspace{1cm}}$

6.

$34 + 4 = \underline{\hspace{1cm}}$ $34 + 6 = \underline{\hspace{1cm}}$ $34 + 9 = \underline{\hspace{1cm}}$

7.

$64 + 4 = \underline{\hspace{1cm}}$ $64 + 6 = \underline{\hspace{1cm}}$ $64 + 9 = \underline{\hspace{1cm}}$

8.

$74 + 4 = \underline{\hspace{1cm}}$ $74 + 6 = \underline{\hspace{1cm}}$ $74 + 9 = \underline{\hspace{1cm}}$

Number and Operations in Base Ten 2-5

☐ Count on to add.

q.

$48 + 1 = $ _____ $58 + 1 = $ _____ $68 + 1 = $ _____

10.

$48 + 2 = $ _____ $58 + 2 = $ _____ $68 + 2 = $ _____

11.

$48 + 3 = $ _____ $58 + 3 = $ _____ $68 + 3 = $ _____

12.

$57 + 2 = $ _____ $67 + 2 = $ _____ $77 + 2 = $ _____

13.

$57 + 3 = $ _____ $67 + 3 = $ _____ $77 + 3 = $ _____

14.

$57 + 5 = $ _____ $67 + 5 = $ _____ $77 + 5 = $ _____

15.

$65 + 4 = $ _____ $75 + 4 = $ _____ $85 + 4 = $ _____

16.

$65 + 5 = $ _____ $75 + 5 = $ _____ $85 + 5 = $ _____

17.

$65 + 9 = $ _____ $75 + 9 = $ _____ $85 + 9 = $ _____

Number and Operations in Base Ten 2-5

NBT2-6 Adding 10

☐ Circle the next 10 numbers.
☐ Add 10.

1.

1	2	3	4	⑤	⑥	7	⑧	⑨	⑩
⑪	⑫	⑬	⑭	15	16	17	18	19	20

$4 + 10 = \underline{\ 14\ }$

2.

11	12	13	14	15	16	17	18	19	20
21	22	23	24	25	26	27	28	29	30

$19 + 10 = \underline{\qquad}$

3.

31	32	33	34	35	36	37	38	39	40
41	42	43	44	45	46	47	48	49	50

$38 + 10 = \underline{\qquad}$

4.

81	82	83	84	85	86	87	88	89	90
91	92	93	94	95	96	97	98	99	100

$90 + 10 = \underline{\qquad}$

☐ Add 10 by moving down a row.

5.

1	2	3	4	5	6	7	8	9	10
11	12	13	14	15	16	17	18	19	20

$3 + 10 = \underline{\qquad}$

$7 + 10 = \underline{\qquad}$

$9 + 10 = \underline{\qquad}$

Does the picture show adding ten?
☐ Circle **Yes** or **No**.

6.

51	52	53	**54**	55	56	57	58	59	60
61	62	63	64	**65**	66	67	68	69	70

Yes ⠀ (No)

7.

71	72	73	74	75	76	77	78	79	**80**
81	82	83	84	85	86	87	88	89	**90**

Yes ⠀ No

8.

31	32	33	34	35	36	37	38	**39**	40
41	42	43	44	45	46	47	48	49	50

Yes ⠀ No

☐ Move down a row to add 10.

11	12	13	14	15	16	17	18	19	20
21	22	23	24	25	26	27	28	29	30
31	32	33	34	35	36	37	38	39	40

9.

17 + 10 = _____

10.

25 + 10 = _____

11.

18 + 10 = _____

12.

11 + 10 = _____

13.

23 + 10 = _____

14.

30 + 10 = _____

NBT2-7 Counting by Tens

These numbers are **multiples of 10**.

10, 20, 30, 40, 50, 60, 70, 80, 90

☐ Fill in the missing multiples of 10.

1.
10, __20__ , __30__ , __40__ , 50

2.
40, _____, _____, _____, 80

3.
40, _____, _____, _____, _____, 90

4.
20, _____, _____, _____, _____, 70

5.
10, _____, _____, _____, _____, _____, _____, _____, 90

☐ Circle the multiples of 10.

6.
48, 49, (50), 51, 52, 53, 54

7.
85, 86, 87, 88, 89, 90, 91

8.
17, 18, 19, 20, 21, 22, 23, 24, 25, 26, 27, 28, 29, 30, 31, 32

☐ Circle the next multiple of 10.

9.
12   10, (20), 30, 40

10.
38   40, 50, 60, 70

11.
79   50, 60, 70, 80

12.
45   30, 40, 50, 60

13.
67   60, 70, 80, 90

14.
53   30, 40, 50, 60

NBT2-8 Using Base Ten Blocks to Add (No Regrouping)

○ Write the number of tens and ones the addition makes.

○ Add.

1.

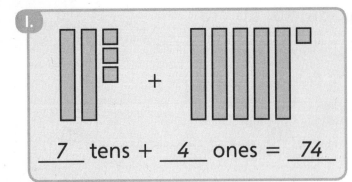

__7__ tens + __4__ ones = __74__

2.

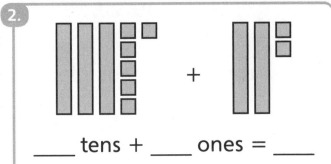

____ tens + ____ ones = ____

3.

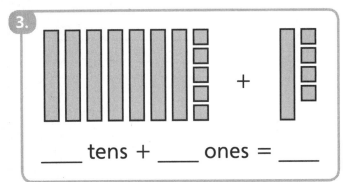

____ tens + ____ ones = ____

4.

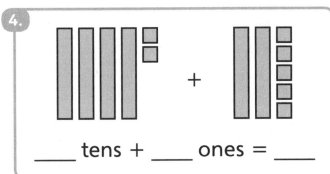

____ tens + ____ ones = ____

○ Write the numbers that the blocks show.

○ Use the blocks to help add.

5.

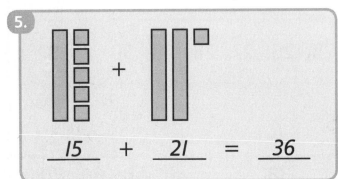

__15__ + __21__ = __36__

6.

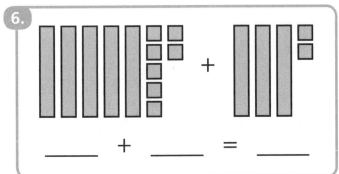

____ + ____ = ____

7.

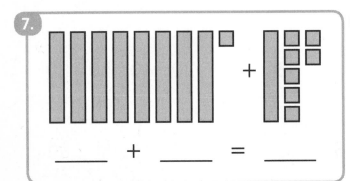

____ + ____ = ____

8.

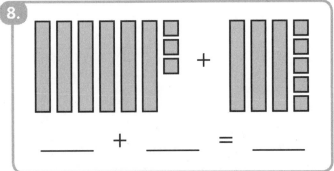

____ + ____ = ____

☐ Draw tens and ones to show the numbers.
☐ Use the drawings to help add.

9.

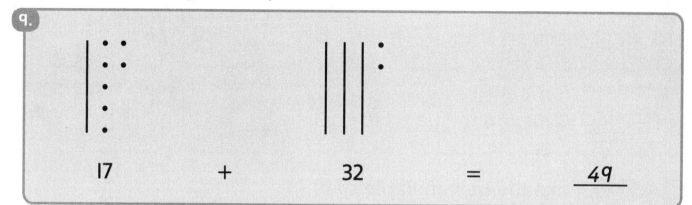

17   +   32   =   49

10.

25   +   41   =   _____

11.

54   +   35   =   _____

12.

72   +   16   =   _____

NBT2-9 Making 10 to Add (1-Digit to 2-Digit)

☐ Add to make 10.

1. 8 + _2_ = 10

2. 7 + ___ = 10

3. 5 + ___ = 10

4. 6 + ___ = 10

5. 1 + ___ = 10

6. 4 + ___ = 10

7. 3 + ___ = 10

8. 2 + ___ = 10

☐ Add to make the next multiple of 10.

9. 7 + _3_ = 10   47 + _3_ = 50   87 + _3_ = 90

10. 2 + ___ = 10   32 + ___ = 40   72 + ___ = 80

11. 6 + ___ = 10   26 + ___ = 30   56 + ___ = 60

12. 1 + ___ = 10   11 + ___ = 20   61 + ___ = 70

☐ Add to make the next multiple of 10.

13. 28 + _2_ = 30

14. 34 + ___ = 40

15. 49 + ___ = 50

16. 21 + ___ = 30

17. 67 + ___ = 70

18. 52 + ___ = 60

19. 75 + ___ = 80

20. 86 + ___ = 90

21. 33 + ___ = 40

Fill in the blank so that the first two numbers add to 10.

22.
8 + 5

8 + ___2___ + 3

23.
6 + 9

6 + ____ + 5

24.
3 + 9

3 + ____ + 2

25.
7 + 8

7 + ____ + 5

26.
9 + 4

9 + ____ + 3

27.
5 + 7

5 + ____ + 2

28.
4 + 8

4 + ____ + 2

29.
9 + 9

9 + ____ + 8

Circle the one that is easiest to add.

30.

34 + 7

34 + 2 + 5

34 + 7

34 + 6 + 1

34 + 7

34 + 3 + 4

31.

85 + 9

85 + 8 + 1

85 + 9

85 + 3 + 6

85 + 9

85 + 5 + 4

32.

67 + 8

67 + 3 + 5

67 + 8

67 + 7 + 1

67 + 8

67 + 2 + 6

☐ Fill in the blanks so that the first two numbers add to a multiple of 10.

33.

17 + 4

17 + _3_ + _1_

17 + 5

17 + ___ + ___

17 + 8

17 + ___ + ___

34.

24 + 7

24 + ___ + ___

24 + 8

24 + ___ + ___

24 + 9

24 + ___ + ___

35.

58 + 3

58 + ___ + ___

58 + 6

58 + ___ + ___

58 + 8

58 + ___ + ___

36.

78 + 5

78 + _2_ + _3_

37.

26 + 9

26 + ___ + ___

38.

35 + 6

35 + ___ + ___

39.

67 + 4

67 + ___ + ___

40.

38 + 5

38 + ___ + ___

41.

59 + 3

59 + ___ + ___

42.

29 + 7

29 + ___ + ___

43.

43 + 8

43 + ___ + ___

44.

76 + 6

76 + ___ + ___

☐ Fill in the blanks so that the first two numbers add to a multiple of 10.

☐ Add to make the next multiple of 10.

45.

53 + 8

53 + _7_ + _1_

_60_ + _1_

46.

29 + 3

29 + ___ + ___

___ + ___

47.

37 + 6

37 + ___ + ___

___ + ___

48.

18 + 5

18 + ___ + ___

___ + ___

49.

77 + 4

77 + ___ + ___

___ + ___

50.

86 + 9

86 + ___ + ___

___ + ___

☐ Add.

51.

29 + 7

29 + _1_ + _6_

_30_ + _6_

_36_

52.

36 + 6

36 + ___ + ___

___ + ___

___

53.

14 + 8

14 + ___ + ___

___ + ___

___

54.

73 + 9

73 + ___ + ___

___ + ___

___

55.

44 + 8

44 + ___ + ___

___ + ___

___

56.

85 + 9

85 + ___ + ___

___ + ___

___

NBT2-10 Using Base Ten Blocks to Add (Regrouping)

- ☐ Group 10 ones to make a ten.
- ☐ Write the number of tens.
- ☐ How many ones are left?
- ☐ Add.

1.

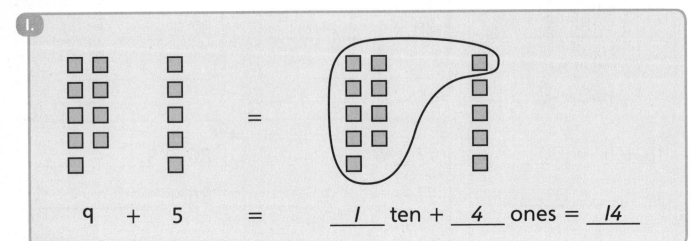

9 + 5 = ___1___ ten + ___4___ ones = ___14___

2.

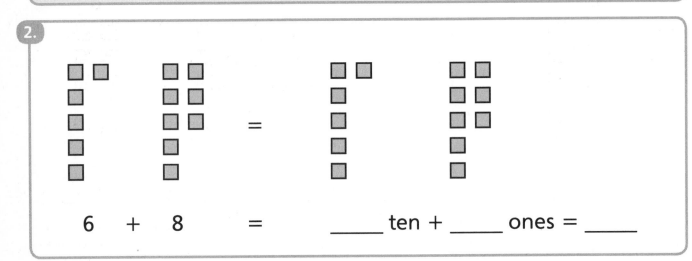

6 + 8 = _____ ten + _____ ones = _____

3.

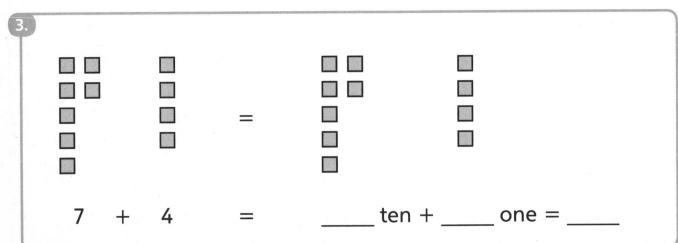

7 + 4 = _____ ten + _____ one = _____

☐ Group 10 ones to make a ten.

☐ Add the tens. Then add on the ones that are left.

4.

28 + 6

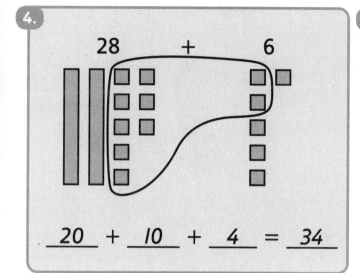

_20_ + _10_ + _4_ = _34_

5.

18 + 3

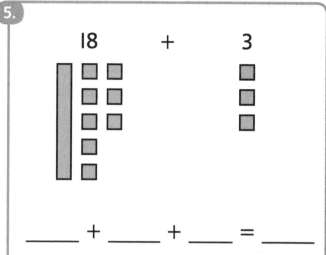

____ + ____ + ____ = ____

6.

35 + 7

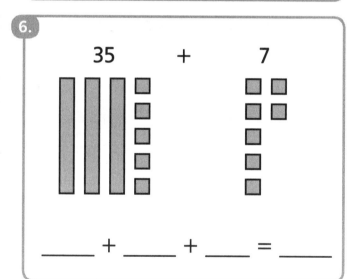

____ + ____ + ____ = ____

7.

43 + 8

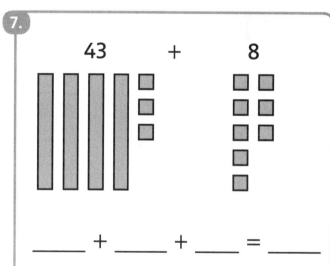

____ + ____ + ____ = ____

8.

14 + 9

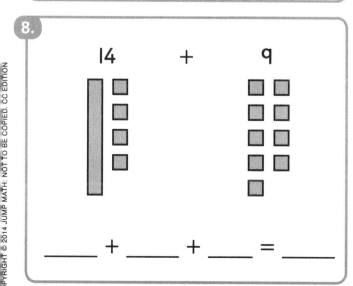

____ + ____ + ____ = ____

9.

55 + 6

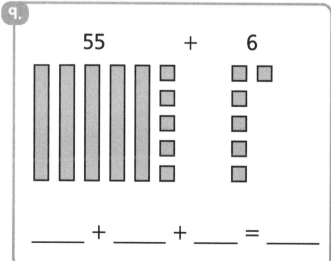

____ + ____ + ____ = ____

NBT2-II Making 10 to Add (2-Digit to 2-Digit)

☐ Add.

1.
2 + 3 = __5__

20 + 30 = __50__

2.
1 + 5 = _____

10 + 50 = _____

3.
3 + 4 = _____

30 + 40 = _____

4.
5 + 1 + 1 + 1 = _____

50 + 10 + 10 + 10 = _____

5.
2 + 3 + 2 + 1 = _____

20 + 30 + 20 + 10 = _____

☐ Write the second number in expanded form.

6.
30 + 15

= 30 + __10__ + __5__

7.
50 + 19

= 50 + ____ + ____

8.
40 + 17

= 40 + ____ + ____

☐ Write the second number in expanded form.
☐ Add the tens.

9.
60 + 11

= 60 + __10__ + __1__

= __70__ + __1__

10.
70 + 13

= 70 + ____ + ____

= ____ + ____

11.
80 + 14

= 80 + ____ + ____

= ____ + ____

☐ Write the second number in expanded form.
☐ Add the tens and ones.

12.
20 + 14

= 20 + __10__ + __4__

= __30__ + __4__

= __34__

13.
50 + 18

= 50 + ____ + ____

= ____ + ____

= ____

14.
70 + 12

= 70 + ____ + ____

= ____ + ____

= ____

Number and Operations in Base Ten 2-II

☐ Add by separating the tens and ones.

15.

23	=	20	+	3
+ 34	=	30	+	4
57	←	50	+	7

16.

34	=	30	+	4
+ 15	=	10	+	5
☐	←	40	+	9

17.

27	=	20	+	☐
+ 22	=	20	+	☐
☐	←	40	+	☐

18.

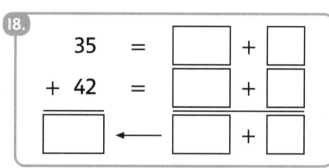

35	=	☐	+	☐
+ 42	=	☐	+	☐
☐	←	☐	+	☐

19.

15	=	☐	+	☐
+ 23	=	☐	+	☐
☐	←	☐	+	☐

20.

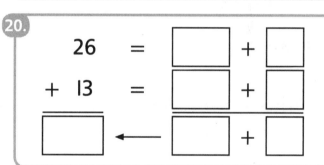

26	=	☐	+	☐
+ 13	=	☐	+	☐
☐	←	☐	+	☐

21.

34	=	☐	+	☐
+ 54	=	☐	+	☐
☐	←	☐	+	☐

22.

26	=	☐	+	☐
+ 33	=	☐	+	☐
☐	←	☐	+	☐

23.

34	=	☐	+	☐
13	=	☐	+	☐
+ 52	=	☐	+	☐
☐	←	☐	+	☐

24.

17	=	☐	+	☐
20	=	☐	+	☐
+ 61	=	☐	+	☐
☐	←	☐	+	☐

☐ Add by using a tens and ones chart.

25.

35
+ 32
─────
67 ←

Tens	Ones
3	5
3	2
6	7

26.

24
+ 41
─────
[ ] ←

Tens	Ones
2	4
4	1

27.

46
+ 31
─────
[ ] ←

Tens	Ones

28.

43
+ 23
─────
[ ] ←

Tens	Ones

29.

27
+ 21
+ 51
─────
[ ] ←

Tens	Ones

30.

31
+ 42
+ 14
─────
[ ] ←

Tens	Ones

31. BONUS

37	63	25	31	54	23
+ 22	+ 16	+ 34	+ 62	+ 34	+ 43

32.

Clara collects stickers. She has
16 bird stickers and 22 animal stickers.
How many stickers does Clara have?

☐ Circle 10 ones to make a ten.

☐ **Regroup** in the next row.

33.

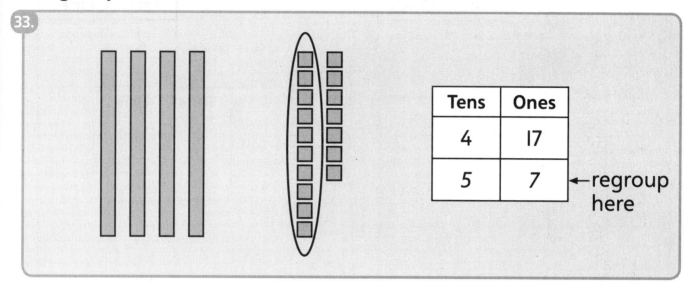

Tens	Ones
4	17
5	7

←regroup here

34.

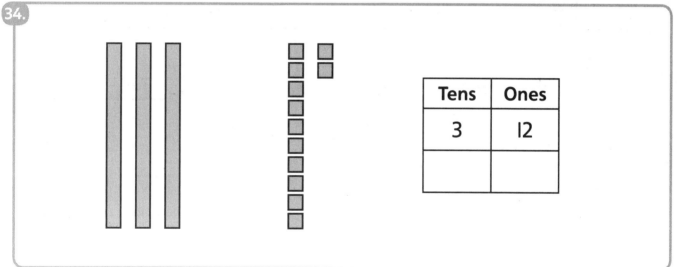

Tens	Ones
3	12

35.

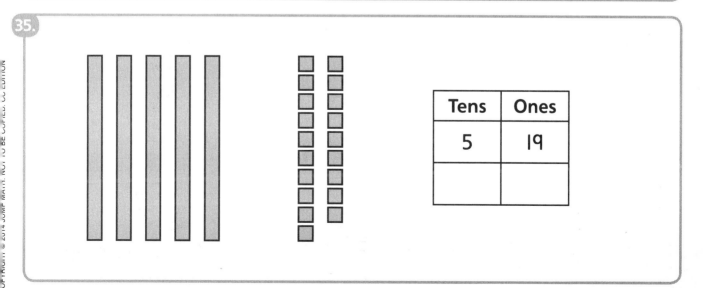

Tens	Ones
5	19

☐ Circle 10 ones to make a ten.
☐ **Regroup** in the next row.

36.

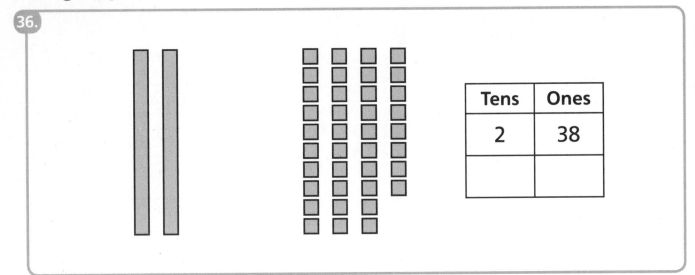

Tens	Ones
2	38

37.

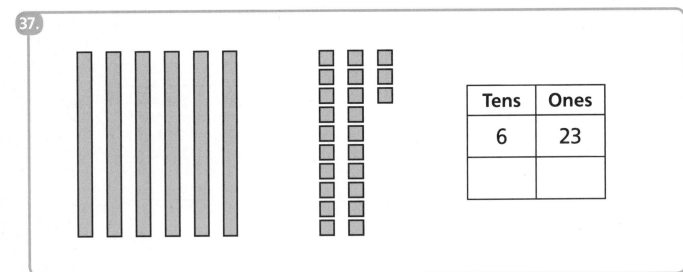

Tens	Ones
6	23

☐ Regroup in the next row.

38.

Tens	Ones
5	31

39.

Tens	Ones
7	25

40.

Tens	Ones
6	24

Number and Operations in Base Ten 2-11

☐ Add the tens and the ones.
☐ Regroup in the next row.
☐ Write the answer.

41.

Tens	Ones
1	6
5	5
6	11
7	1

16
+ 55
——
71

42.

Tens	Ones
1	2
2	9

12
+ 29
——

43.

Tens	Ones
2	5
3	8

25
+ 38
——

44.

Tens	Ones
5	7
2	6

57
+ 26
——

45.

Tens	Ones
1	6
3	4
2	8

16
34
+ 28
——

46.

Tens	Ones
2	3
5	2
1	6

23
52
+ 16
——

NBT2-12 Using Place Value to Add (No Regrouping)

☐ Add the ones.

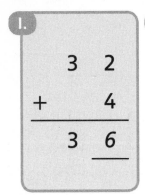

1.
```
    3  2
+      4
─────────
    3  6
```

2.
```
    7  3
+      5
─────────
    7  __
```

3.
```
    2  0
+      8
─────────
    2  __
```

4.
```
    4  1
+      6
─────────
    4  __
```

5.
```
    5  5
+      2
─────────
    5  __
```

6.
```
    8  2
+      3
─────────
    8  __
```

7.
```
    6  1
+      7
─────────
    6  __
```

8.
```
    1  0
+      9
─────────
    1  __
```

9.
```
    9  4
+      2
─────────
    9  __
```

10.
```
    3  8
+      1
─────────
    3  __
```

☐ Add the ones.
☐ Add the tens.

11.
```
    4  1
+  2  4
─────────
   __ __
```

12.
```
    3  2
+  3  1
─────────
   __ __
```

13.
```
    5  1
+  2  8
─────────
   __ __
```

14.
```
    4  4
+  4  0
─────────
   __ __
```

15.
```
    2  7
+  1  2
─────────
   __ __
```

16.
```
    6  1
+  1  8
─────────
   __ __
```

17.
```
    7  0
+  2  4
─────────
   __ __
```

18.
```
    8  9
+  1  0
─────────
   __ __
```

19.
```
    5  6
+  4  3
─────────
   __ __
```

20.
```
    7  3
+  1  5
─────────
   __ __
```

NBT2-13 Using Place Value to Add (Regrouping)

◯ Group 10 ones.
◯ In the ▢ box, write 1 to show the group of 10 ones.
◯ In the ▢ box, write the number of ones left.

1.
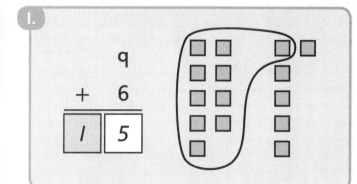

$$\begin{array}{r} 9 \\ + \ 6 \\ \hline \end{array}$$
[1][5]

2.
$$\begin{array}{r} 8 \\ + \ 3 \\ \hline \end{array}$$
[ ][ ]
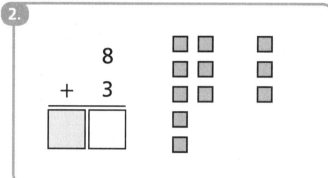

3.
$$\begin{array}{r} 5 \\ + \ 6 \\ \hline \end{array}$$
[ ][ ]
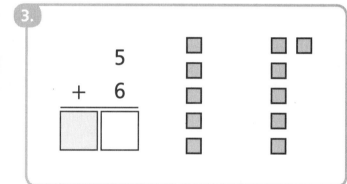

4.
$$\begin{array}{r} 7 \\ + \ 5 \\ \hline \end{array}$$
[ ][ ]
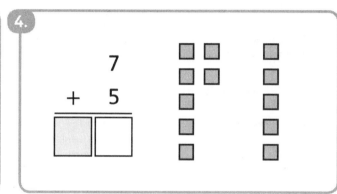

5.
$$\begin{array}{r} 8 \\ + \ 6 \\ \hline \end{array}$$
[ ][ ]
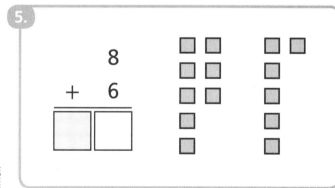

6.
$$\begin{array}{r} 7 \\ + \ 9 \\ \hline \end{array}$$
[ ][ ]
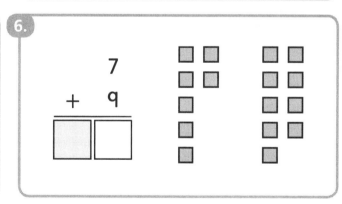

7.
$$\begin{array}{r} 5 \\ + \ 8 \\ \hline \end{array}$$
[ ][ ]
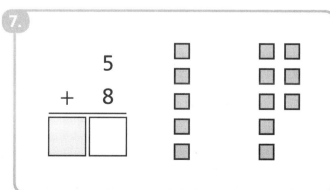

8.
$$\begin{array}{r} 9 \\ + \ 9 \\ \hline \end{array}$$
[ ][ ]

☐ Group 10 ones.
☐ Above the addition, write the number of tens.
☐ Below the addition, write the number of ones.

9.

18   +   25

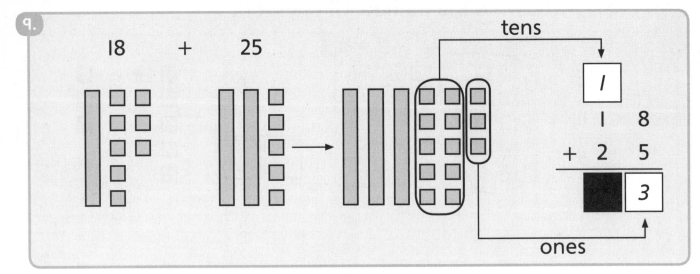

10.

39   +   13

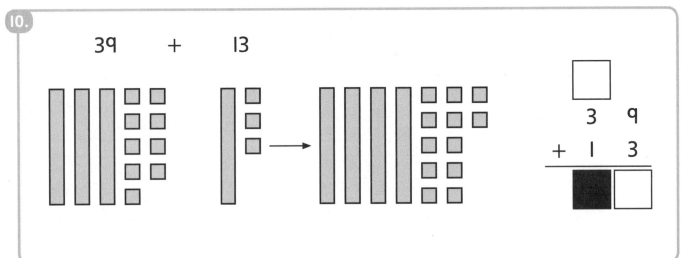

11.

47   +   16

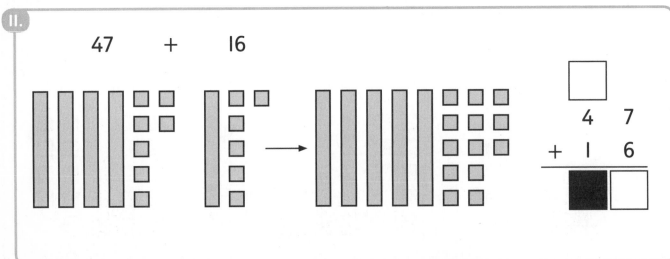

Number and Operations in Base Ten 2-13

⬜ Add the ones using the blanks at the top.
⬜ Above the tens, write 1 to show 10 ones.
⬜ Below the ones, write the number of ones left.

12.

$5 + 9 =$ __1__ __4__

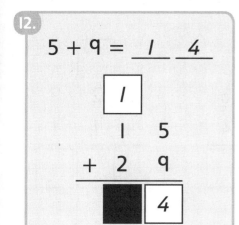

	1	
	1	5
+	2	9
	■	4

13.

$4 + 8 =$ __1__ __2__

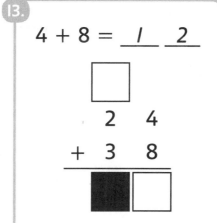

	2	4
+	3	8
	■	☐

14.

$6 + 4 =$ __1__ __0__

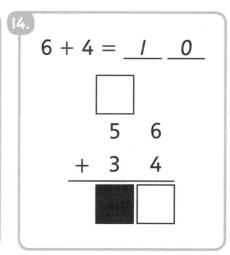

	5	6
+	3	4
	■	☐

15.

$7 + 5 =$ ___ ___

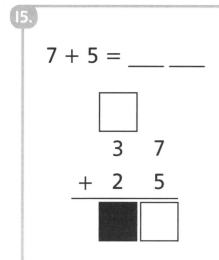

	3	7
+	2	5
	■	☐

16.

$6 + 9 =$ ___ ___

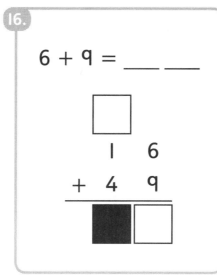

	1	6
+	4	9
	■	☐

17.

___ $+$ ___ $=$ ___ ___

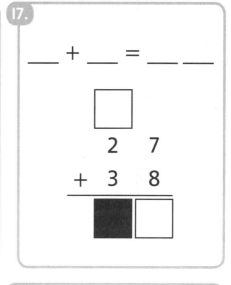

	2	7
+	3	8
	■	☐

18.

___ $+$ ___ $=$ ___ ___

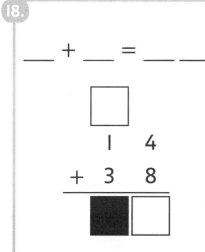

	1	4
+	3	8
	■	☐

19.

___ $+$ ___ $=$ ___ ___

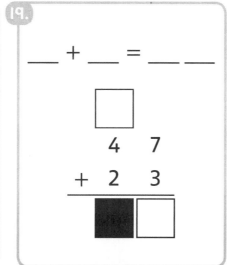

	4	7
+	2	3
	■	☐

20.

___ $+$ ___ $=$ ___ ___

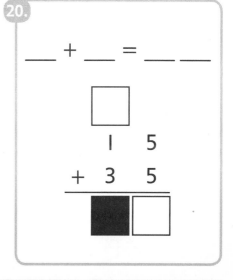

	1	5
+	3	5
	■	☐

☐ Add the ones first.

☐ Then add the tens to find the total.

21.

```
    1
    1   5
+   2   9
─────────
    4   4
```

22.
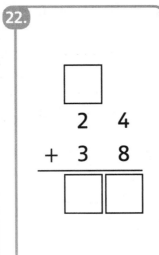
```
    2   4
+   3   8
─────────
```

23.
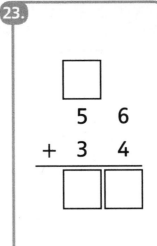
```
    5   6
+   3   4
─────────
```

24.
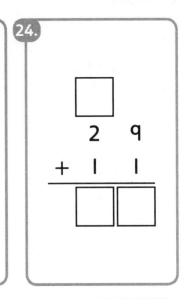
```
    2   9
+   1   1
─────────
```

25.
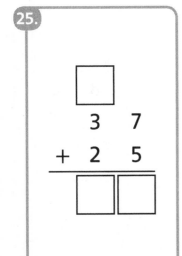
```
    3   7
+   2   5
─────────
```

26.
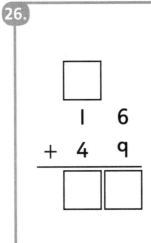
```
    1   6
+   4   9
─────────
```

27.
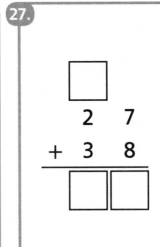
```
    2   7
+   3   8
─────────
```

28.
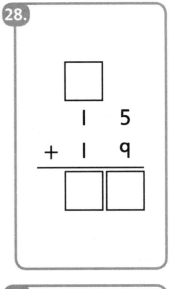
```
    1   5
+   1   9
─────────
```

29.
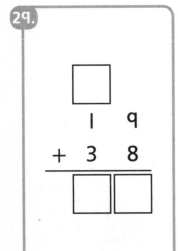
```
    1   9
+   3   8
─────────
```

30.
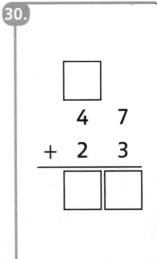
```
    4   7
+   2   3
─────────
```

31.
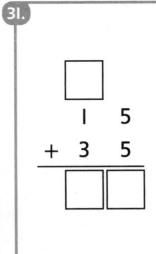
```
    1   5
+   3   5
─────────
```

32.
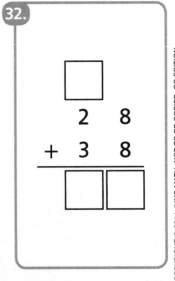
```
    2   8
+   3   8
─────────
```

☐ Regroup only when you need to.
☐ Add.

33.
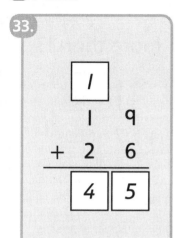
```
    1
  1   9
+ 2   6
─────────
  4   5
```

34.
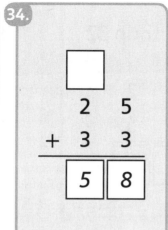
```
    ☐
  2   5
+ 3   3
─────────
  5   8
```

35.
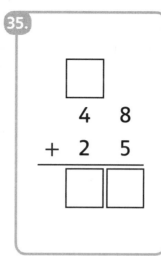
```
    ☐
  4   8
+ 2   5
─────────
  ☐   ☐
```

36.
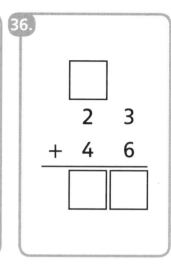
```
    ☐
  2   3
+ 4   6
─────────
  ☐   ☐
```

37.
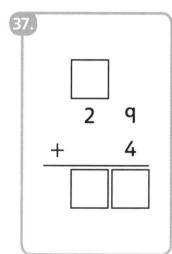
```
    ☐
  2   9
+     4
─────────
  ☐   ☐
```

38.
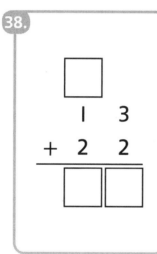
```
    ☐
  1   3
+ 2   2
─────────
  ☐   ☐
```

39.
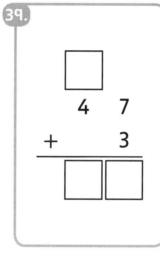
```
    ☐
  4   7
+     3
─────────
  ☐   ☐
```

40.
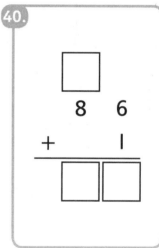
```
    ☐
  8   6
+     1
─────────
  ☐   ☐
```

Ivan added the tens before the ones.

☐ Circle the answers that are incorrect.

41.
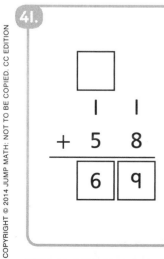
```
    ☐
  1   1
+ 5   8
─────────
  6   9
```

42.
```
    1
  1   7
+ 2   7
─────────
  3   4
```

43.
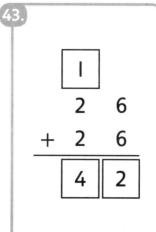
```
    1
  2   6
+ 2   6
─────────
  4   2
```

44.
```
    ☐
  4   3
+ 2   5
─────────
  6   8
```

NBT2-14 Addition Word Problems

☐ Add.

1. 11 more than 24

```
  2 4
+ 1 1
─────
  3 5
```

2. 45 more than 32

```
  3 2
+ 4 5
─────
  [ ] [ ]
```

3. 76 more than 13

```
  1 3
+ 7 6
─────
  [ ] [ ]
```

4. 17 more than 15

```
[1]
  1 5
+ 1 7
─────
  3 2
```

5. 29 more than 36

```
[ ]
  3 6
+ 2 9
─────
  [ ] [ ]
```

6. 48 more than 47

```
[ ]
  4 7
+ 4 8
─────
  [ ] [ ]
```

7. 18 cats

25 more dogs than cats

```
[1]
  1 8
+ 2 5
─────
  4 3  dogs
```

8. 23 boxes

29 more bags than boxes

```
[ ]
  2 3
+ 2 9
─────
  [ ] [ ]  bags
```

9. 47 pencils

36 more pens than pencils

```
[ ]
  4 7
+ 3 6
─────
  [ ] [ ]  pens
```

☐ Add.

10.

15 students were on the bus.

27 more students got on.

How many students were
on the bus altogether?

```
      [ 1 ]
      1   5
  +   2   7
    [ 4 ][ 2 ]
```

11.

Beth made 18 muffins on
Saturday and 24 more on Sunday.

How many did she make altogether?

```
      [   ]
      1   8
  +   2   4
    [   ][   ]
```

12.

29 girls and 33 boys were at the park.

How many children were at the park?

```
      [   ]
      2   9
  +   3   3
    [   ][   ]
```

13.

Ross picked 35 pears in the morning
and 39 more in the afternoon.

How many did he pick altogether?

```
      [   ]
      3   5
  +   3   9
    [   ][   ]
```

☐ Add.

14.

Tina drew 18 circles.

Sal drew 3 more circles than Tina.

How many circles did Sal draw?

$$
\begin{array}{r}
\boxed{1}\ \ \ \\
1\ \ 8 \\
+\ \ \ \ 3 \\
\hline
\boxed{2}\ \boxed{1} \\
\end{array}
$$

15.

Raj pitched the ball 25 times.

Josh pitched the ball 17 more times than Raj.

How many times did Josh pitch the ball?

$$
\begin{array}{r}
\boxed{} \\
+\ \ \ \ \ \\
\hline
\boxed{}\ \boxed{} \\
\end{array}
$$

16.

Rick collected 46 rocks.

May collected 48 more rocks than Rick.

How many rocks did May collect?

$$
\begin{array}{r}
\boxed{} \\
+\ \ \ \ \ \\
\hline
\boxed{}\ \boxed{} \\
\end{array}
$$

17.

Hanna wrote 52 words.

Grace wrote 39 more words than Hanna.

How many words did Grace write?

$$
\begin{array}{r}
\boxed{} \\
+\ \ \ \ \ \\
\hline
\boxed{}\ \boxed{} \\
\end{array}
$$

Number and Operations in Base Ten 2-14

☐ Count on to find the missing number.

1.

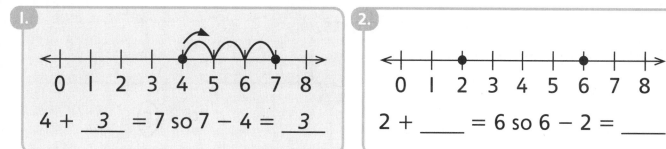

$4 + \underline{\ 3\ } = 7$ so $7 - 4 = \underline{\ 3\ }$

2.

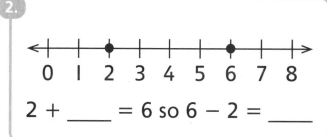

$2 + \underline{\quad} = 6$ so $6 - 2 = \underline{\quad}$

3.

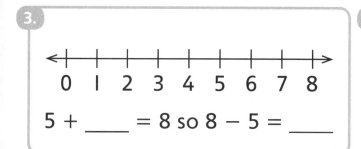

$5 + \underline{\quad} = 8$ so $8 - 5 = \underline{\quad}$

4.

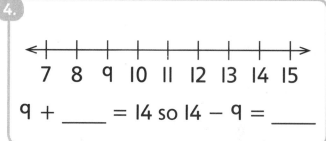

$9 + \underline{\quad} = 14$ so $14 - 9 = \underline{\quad}$

☐ Count on to subtract.

5.

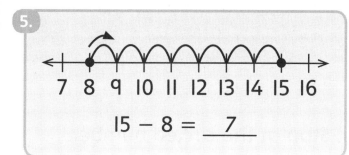

$15 - 8 = \underline{\ 7\ }$

6.

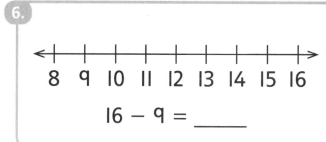

$16 - 9 = \underline{\quad}$

7.

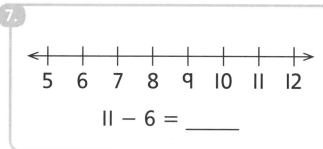

$11 - 6 = \underline{\quad}$

8.

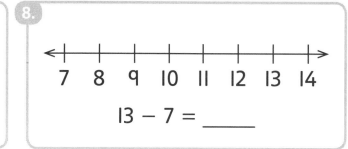

$13 - 7 = \underline{\quad}$

9.

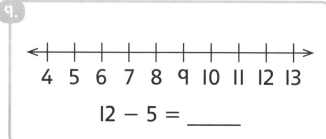

$12 - 5 = \underline{\quad}$

10.

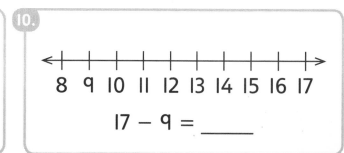

$17 - 9 = \underline{\quad}$

Start counting at the second number.
☐ Circle the second number.
☐ Draw a dot where the count starts.
☐ Write the number.

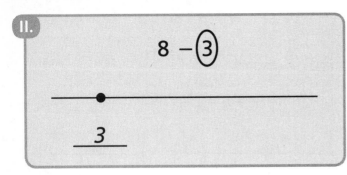

11.
$8 - ③$

_____

3

12.
$9 - ⑤$

_____

_____

13.
$10 - 7$

_____

_____

14.
$11 - 2$

_____

_____

Stop counting at the first number.
☐ Circle the first number.
☐ Add numbers to the number line. Stop at the circled number.

15.
$⑧ - 3$

3 4 5 6 7 8

16.
$⑨ - 5$

5

17.
$10 - 7$

7

18.
$11 - 2$

2

☐ Draw the jumps.
☐ Count the jumps to subtract.

19.

$8 - 3 = \underline{5}$

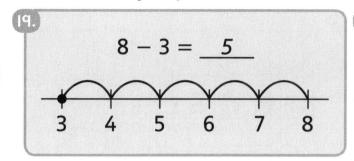

20.

$9 - 5 = \underline{}$

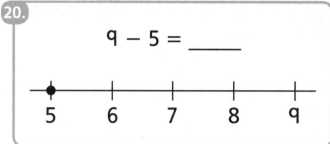

21.

$10 - 7 = \underline{}$

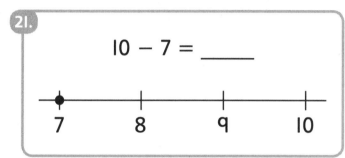

22.

$11 - 2 = \underline{}$

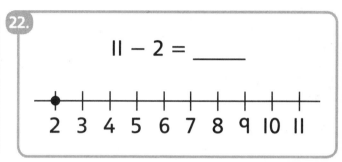

☐ Draw a number line to subtract.

23.

$12 - 5 = \underline{7}$

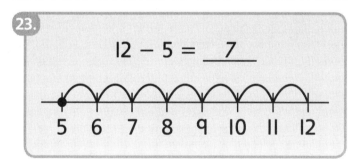

24.

$15 - 9 = \underline{}$

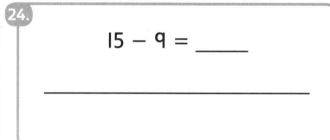

25.

$17 - 9 = \underline{}$

26.

$25 - 22 = \underline{}$

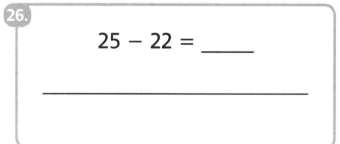

27.

$37 - 31 = \underline{}$

28.

$84 - 77 = \underline{}$

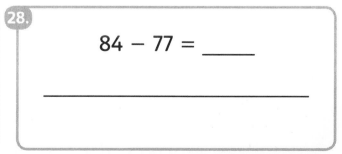

NBT2-16 Subtracting to Make 10 and Subtracting 10

⬚ Color dots to show the addition.
⬚ Subtract.

1.

$$10 + 4 = 14$$

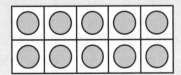 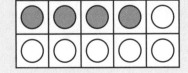

$$14 - 4 = \underline{\ \ 10\ \ }$$ $$14 - 10 = \underline{\ \ 4\ \ }$$

2.

$$10 + 1 = 11$$

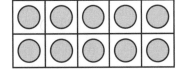 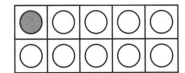

$$11 - 1 = \underline{\ \ \ \ \ }$$ $$11 - 10 = \underline{\ \ \ \ \ }$$

3.

$$10 + 5 = 15$$

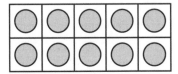 

$$15 - 5 = \underline{\ \ \ \ \ }$$ $$15 - 10 = \underline{\ \ \ \ \ }$$

4.

$$10 + 6 = 16$$

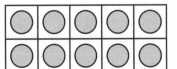 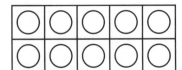

$$16 - 6 = \underline{\ \ \ \ \ }$$ $$16 - 10 = \underline{\ \ \ \ \ }$$

☐ Subtract.

5.
$10 + 3 = 13$

$13 - 3 = \underline{\quad 10 \quad}$

$13 - 10 = \underline{\quad 3 \quad}$

6.
$10 + 2 = 12$

$12 - 2 = \underline{\qquad}$

$12 - 10 = \underline{\qquad}$

7.
$10 + 9 = 19$

$19 - 9 = \underline{\qquad}$

$19 - 10 = \underline{\qquad}$

8.
$16 - 6 = \underline{\qquad}$

9.
$11 - 1 = \underline{\qquad}$

10.
$18 - 8 = \underline{\qquad}$

11.
$17 - 7 = \underline{\qquad}$

12.
$15 - 5 = \underline{\qquad}$

13.
$13 - 10 = \underline{\qquad}$

14.
$11 - 10 = \underline{\qquad}$

15.
$14 - 10 = \underline{\qquad}$

16.
$19 - 10 = \underline{\qquad}$

17.
$15 - 10 = \underline{\qquad}$

18.
$14 - 4 = \underline{\qquad}$

19.
$17 - 10 = \underline{\qquad}$

20.
$18 - 10 = \underline{\qquad}$

21.
$13 - 3 = \underline{\qquad}$

22.
$16 - 10 = \underline{\qquad}$

23. BONUS
$25 - 5 = \underline{\qquad}$

24. BONUS
$37 - 7 = \underline{\qquad}$

25. BONUS
$92 - 2 = \underline{\qquad}$

26. BONUS
$23 - 20 = \underline{\qquad}$

27. BONUS
$47 - 40 = \underline{\qquad}$

28. BONUS
$89 - 80 = \underline{\qquad}$

NBT2-17 Subtracting a Multiple of 10

☐ Subtract 10 by taking away a tens block.

1.
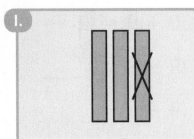
30 − 10 = __20__

2.
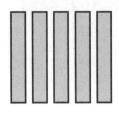
50 − 10 = _____

3.
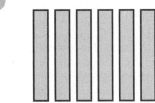
80 − 10 = _____

☐ Subtract by taking away tens blocks.

4.
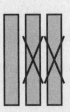
30 − 20 = __10__

5.
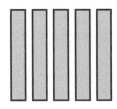
50 − 20 = _____

6.
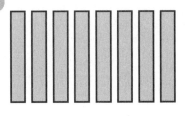
60 − 30 = _____

7.
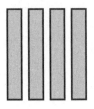
40 − 20 = _____

8.
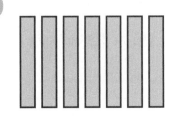
80 − 30 = _____

9.
70 − 50 = _____

10.
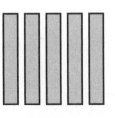
50 − 30 = _____

11.
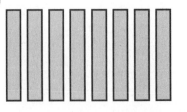
80 − 40 = _____

12.
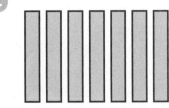
70 − 40 = _____

☐ Subtract 10 by taking away a tens block.

13.

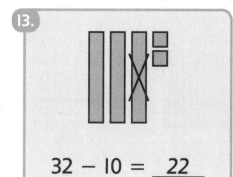

$32 - 10 = \underline{\quad 22 \quad}$

14.

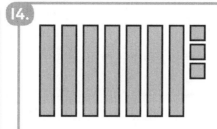

$73 - 10 = \underline{\qquad}$

15.

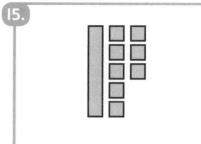

$18 - 10 = \underline{\qquad}$

☐ Subtract by taking away tens blocks.

16.

$65 - 30 = \underline{\quad 35 \quad}$

17.

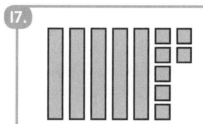

$57 - 20 = \underline{\qquad}$

18.

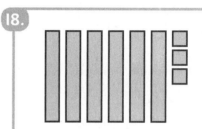

$63 - 40 = \underline{\qquad}$

19.

$52 - 30 = \underline{\qquad}$

20.

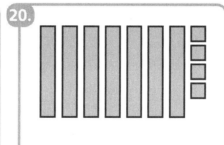

$74 - 60 = \underline{\qquad}$

21.

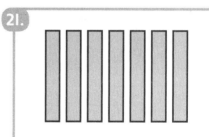

$70 - 50 = \underline{\qquad}$

22.

$41 - 40 = \underline{\qquad}$

23.

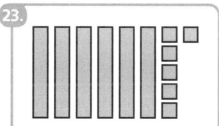

$66 - 50 = \underline{\qquad}$

24.

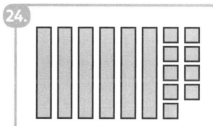

$69 - 40 = \underline{\qquad}$

Number and Operations in Base Ten 2-17

☐ Subtract by taking away tens blocks.

25.
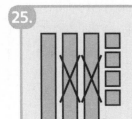

	3	4
−	2	0
	1	4

26.
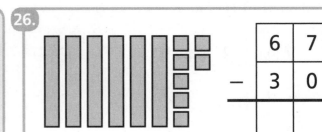

	6	7
−	3	0

27.
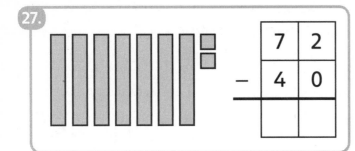

	7	2
−	4	0

28.
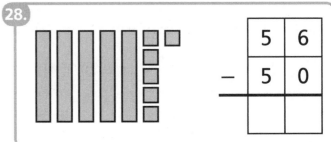

	5	6
−	5	0

☐ Subtract.

29.

	8	3
−	1	0
	7	3

30.

	7	7
−	1	0

31.

	9	2
−	1	0

32.

	3	8
−	2	0

33.

	4	8
−	2	0

34.

	8	6
−	3	0

35.

	9	7
−	4	0

36.

	8	2
−	4	0

37.

	6	1
−	5	0

38.

	9	5
−	8	0

39.

	5	4
−	4	0

40.

	9	9
−	9	0

NBT2-18 Subtracting from a Multiple of 10

◯ Find the missing addend.

1.
7 + __3__ = 10

17 + ____ = 20

57 + ____ = 60

2.
6 + ____ = 10

16 + ____ = 20

56 + ____ = 60

3.
5 + ____ = 10

15 + ____ = 20

55 + ____ = 60

◯ Count on to the next multiple of 10. Then subtract.

4.
10 − 7 = __3__
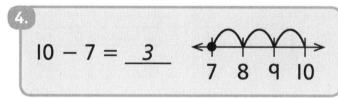

5.
30 − 28 = ____
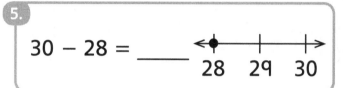

6.
60 − 54 = ____
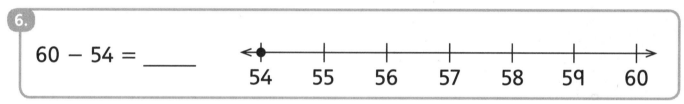

◯ Count the tens to subtract.

7.
70 − 50 = __20__
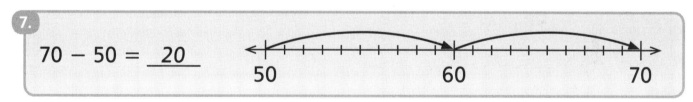

8.
80 − 30 = ____
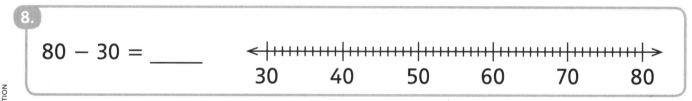

9.
60 − 20 = ____
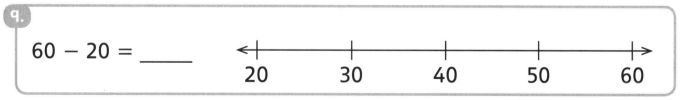

10.
70 − 10 = ____
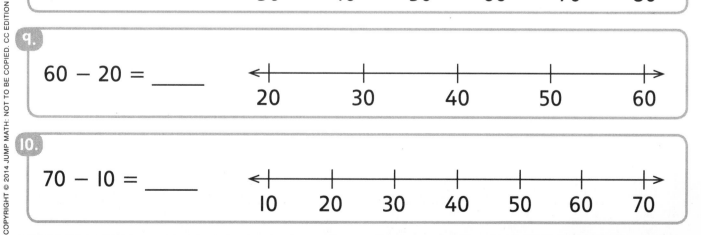

☐ Subtract to find the distances in the picture.

11.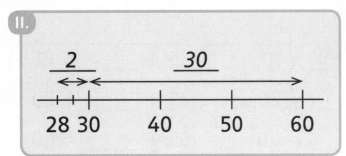

12.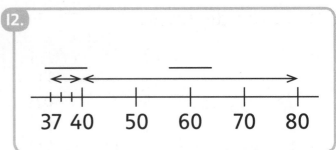

13.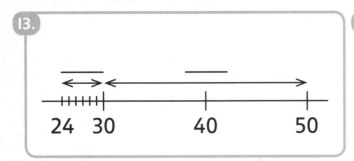

14.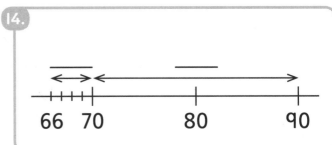

☐ Subtract to find the distances.
☐ Add the distances.

15.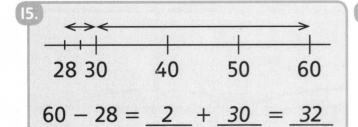

$60 - 28 = \underline{\ 2\ } + \underline{\ 30\ } = \underline{\ 32\ }$

16.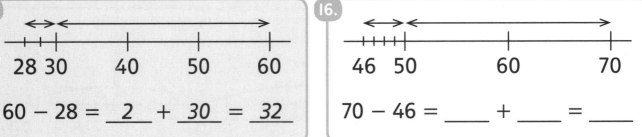

$70 - 46 = \underline{\qquad} + \underline{\qquad} = \underline{\qquad}$

17.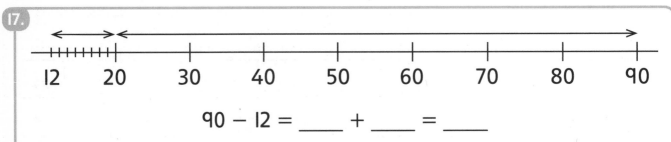

$90 - 12 = \underline{\qquad} + \underline{\qquad} = \underline{\qquad}$

18.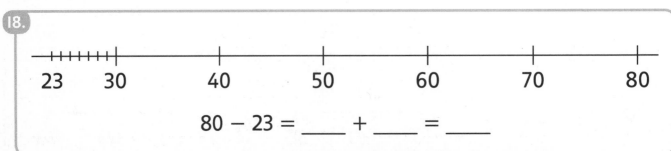

$80 - 23 = \underline{\qquad} + \underline{\qquad} = \underline{\qquad}$

Number and Operations in Base Ten 2-18

☐ Subtract to find the distances.
☐ Add the distances.

19.
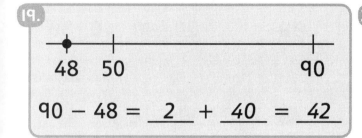
$90 - 48 = \underline{\ 2\ } + \underline{\ 40\ } = \underline{\ 42\ }$

20.
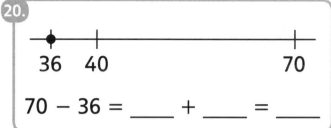
$70 - 36 = \underline{\ \ \ } + \underline{\ \ \ } = \underline{\ \ \ }$

21.
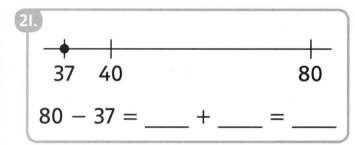
$80 - 37 = \underline{\ \ \ } + \underline{\ \ \ } = \underline{\ \ \ }$

22.
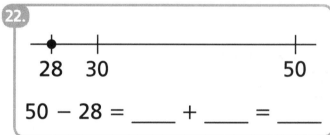
$50 - 28 = \underline{\ \ \ } + \underline{\ \ \ } = \underline{\ \ \ }$

☐ Fill in the missing numbers in the picture.
☐ Find the distances. Then add the distances.

23.
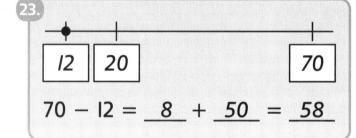
$70 - 12 = \underline{\ 8\ } + \underline{\ 50\ } = \underline{\ 58\ }$

24.
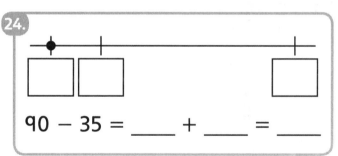
$90 - 35 = \underline{\ \ \ } + \underline{\ \ \ } = \underline{\ \ \ }$

25.
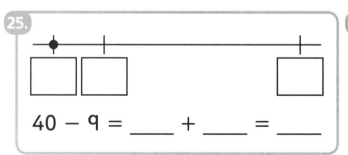
$40 - 9 = \underline{\ \ \ } + \underline{\ \ \ } = \underline{\ \ \ }$

26.
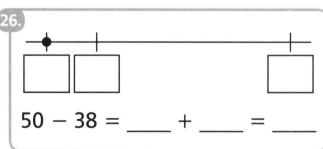
$50 - 38 = \underline{\ \ \ } + \underline{\ \ \ } = \underline{\ \ \ }$

27.
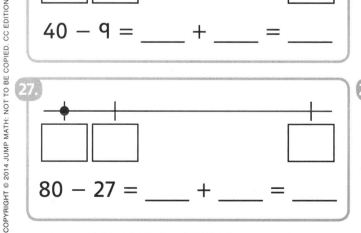
$80 - 27 = \underline{\ \ \ } + \underline{\ \ \ } = \underline{\ \ \ }$

28.
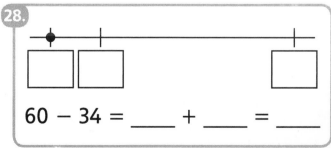
$60 - 34 = \underline{\ \ \ } + \underline{\ \ \ } = \underline{\ \ \ }$

NBT2-19 Subtract by Adding

○ Circle the multiple of 10 that comes before.

1.
32   10, 20, (30), 40

2.
38   20, 30, 40, 50

3.
79   50, 60, 70, 80

4.
45   10, 20, 30, 40

5.
24   10, 20, 30, 40

6.
53   30, 40, 50, 60

7.
12   10, 20, 30, 40

8.
96   60, 70, 80, 90

○ Write the ones digit.

9.
$12 - \underline{\ 2\ } = 10$

10.
$17 - \underline{\ \ \ } = 10$

11.
$25 - \underline{\ \ \ } = 20$

12.
$26 - \underline{\ \ \ } = 20$

13.
$31 - \underline{\ \ \ } = 30$

14.
$54 - \underline{\ \ \ } = 50$

15.
$73 - \underline{\ \ \ } = 70$

16.
$89 - \underline{\ \ \ } = 80$

○ Subtract the ones.

17.
$21 - \underline{\ 1\ } = \underline{\ 20\ }$

18.
$37 - \underline{\ \ \ } = \underline{\ \ \ }$

19.
$46 - \underline{\ \ \ } = \underline{\ \ \ }$

20.
$68 - \underline{\ \ \ } = \underline{\ \ \ }$

21.
$33 - \underline{\ \ \ } = \underline{\ \ \ }$

22.
$92 - \underline{\ \ \ } = \underline{\ \ \ }$

◯ Find the distances in the picture.

23.
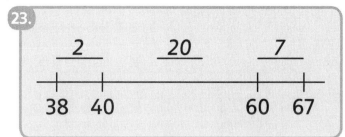

24.
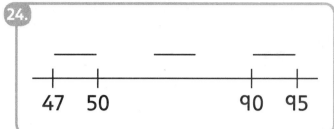

25.
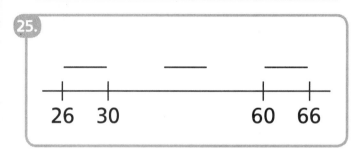

26.
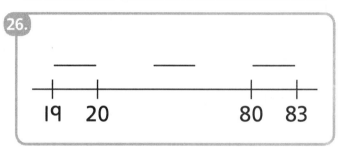

◯ Find the distances in the picture.
◯ Subtract by adding the distances.

27.
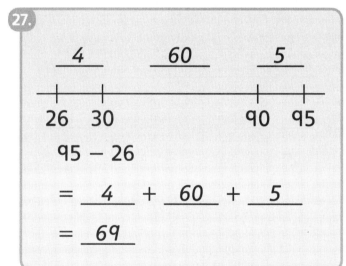

95 − 26

= __4__ + __60__ + __5__

= __69__

28.
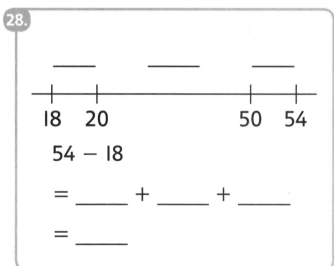

54 − 18

= ____ + ____ + ____

= ____

29.
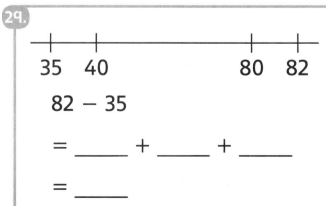

82 − 35

= ____ + ____ + ____

= ____

30.
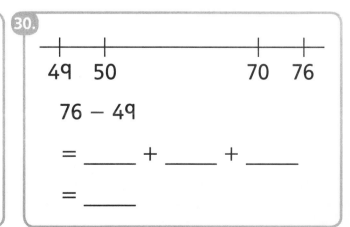

76 − 49

= ____ + ____ + ____

= ____

☐ Fill in the missing numbers in the picture.
☐ Find the distances. Subtract by adding the distances.

31.
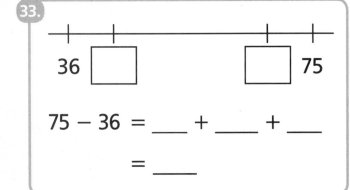
$68 \boxed{70} \qquad \boxed{90}\ 93$

$93 - 68 = \underline{\ 2\ } + \underline{\ 20\ } + \underline{\ 3\ }$

$\qquad\qquad = \underline{\ 25\ }$

32.
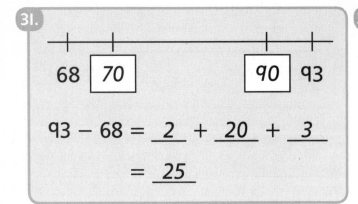
$24 \boxed{30} \qquad \boxed{}\ 56$

$56 - 24 = \underline{\ \ \ } + \underline{\ \ \ } + \underline{\ \ \ }$

$\qquad\qquad = \underline{\ \ \ }$

33.
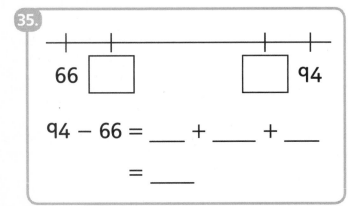
$36 \boxed{} \qquad \boxed{}\ 75$

$75 - 36 = \underline{\ \ \ } + \underline{\ \ \ } + \underline{\ \ \ }$

$\qquad\qquad = \underline{\ \ \ }$

34.
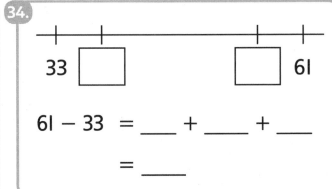
$33 \boxed{} \qquad \boxed{}\ 61$

$61 - 33 = \underline{\ \ \ } + \underline{\ \ \ } + \underline{\ \ \ }$

$\qquad\qquad = \underline{\ \ \ }$

35.
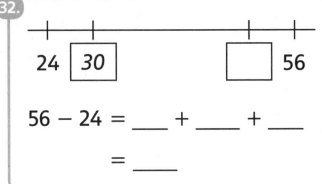
$66 \boxed{} \qquad \boxed{}\ 94$

$94 - 66 = \underline{\ \ \ } + \underline{\ \ \ } + \underline{\ \ \ }$

$\qquad\qquad = \underline{\ \ \ }$

36.
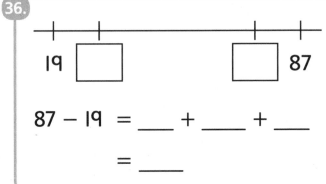
$19 \boxed{} \qquad \boxed{}\ 87$

$87 - 19 = \underline{\ \ \ } + \underline{\ \ \ } + \underline{\ \ \ }$

$\qquad\qquad = \underline{\ \ \ }$

37.
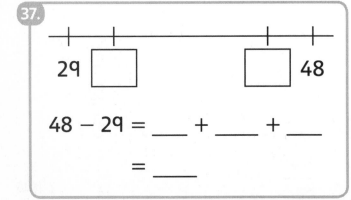
$29 \boxed{} \qquad \boxed{}\ 48$

$48 - 29 = \underline{\ \ \ } + \underline{\ \ \ } + \underline{\ \ \ }$

$\qquad\qquad = \underline{\ \ \ }$

38.
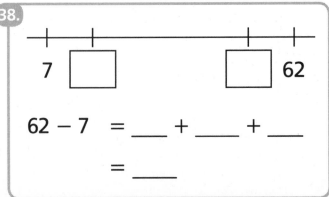
$7 \boxed{} \qquad \boxed{}\ 62$

$62 - 7 = \underline{\ \ \ } + \underline{\ \ \ } + \underline{\ \ \ }$

$\qquad\qquad = \underline{\ \ \ }$

Number and Operations in Base Ten 2-19

NBT2-20 Using Base Ten Blocks to Subtract

☐ Cross out blocks to show the number you take away.
☐ How many blocks are left?

1.

	4	7
−	2	3
	2	4

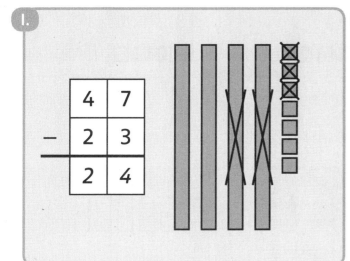

2.

	3	4
−	1	3

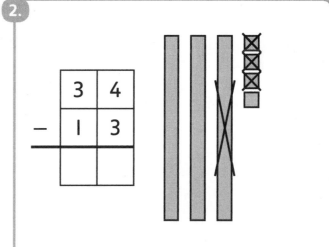

3.

	4	8
−	3	1

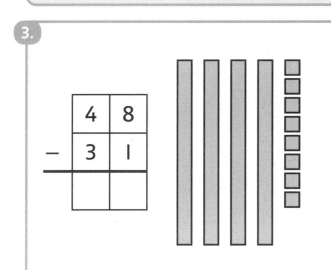

4.

	4	5
−	2	0

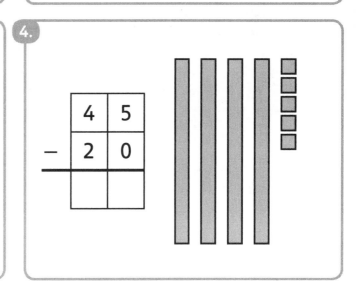

5.

	3	8
−	2	2

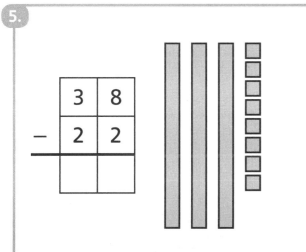

6.

	3	9
−	1	8

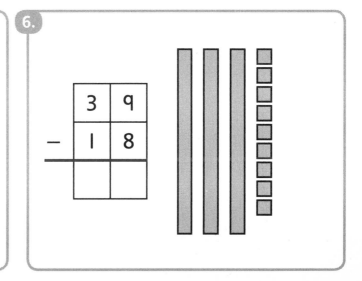

☐ Take apart a tens block for 10 ones.

7.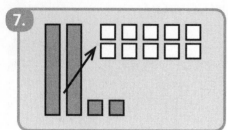

8.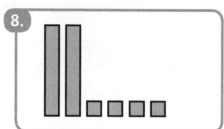

q.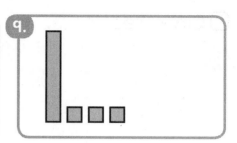

☐ Take apart a tens block.
☐ Subtract.

10.
32
− 19

13
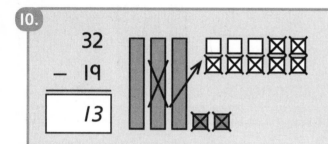

11.
21
− 7

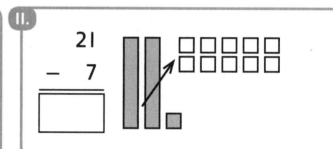

12.
43
− 15

13.
51
− 34

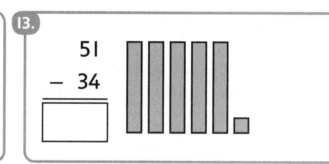

14.
36
− 9

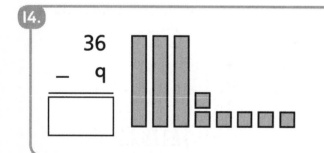

15.
24
− 15

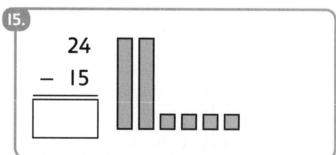

16.
56
− 27

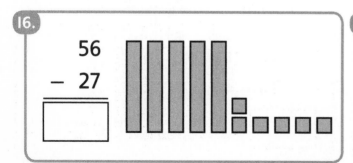

17.
45
− 8

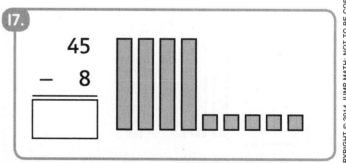

Number and Operations in Base Ten 2-20

Do you need to take apart a tens block to subtract?
☐ Circle **Yes** or **No**.

18.
21
− 18
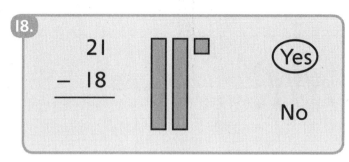
(Yes)
No

19.
39
− 12
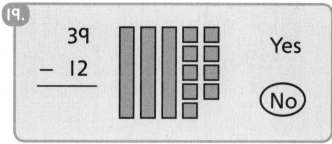
Yes
(No)

20.
33
− 6
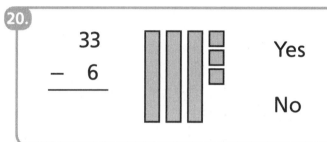
Yes
No

21.
46
− 14
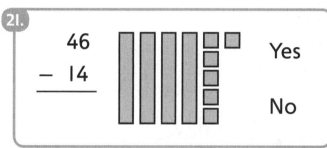
Yes
No

☐ Take apart a tens block only if you need to.
☐ Subtract.

22.
41
− 25
[16]

23.
37
− 12
[ ]
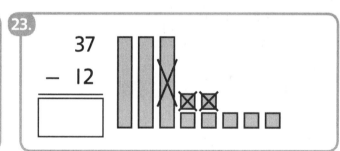

24.
52
− 17
[ ]
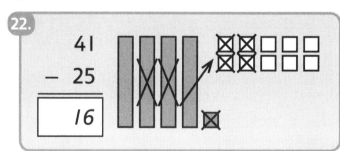

25.
34
− 19
[ ]
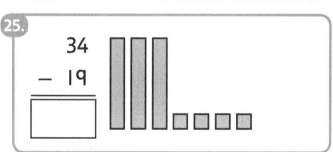

26.
23
− 11
[ ]
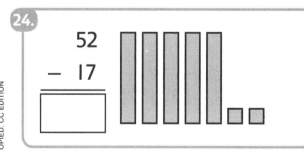

27.
62
− 34
[ ]
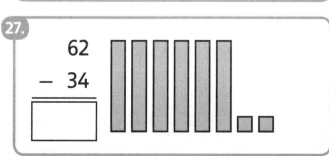

NBT2-21 Using Place Value to Subtract (No Regrouping)

◯ Draw a picture to show the subtraction.
◯ Subtract.

1.
$$\begin{array}{r} 46 \\ -\ 12 \\ \hline 34 \end{array}$$
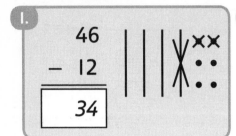

2.
$$\begin{array}{r} 54 \\ -\ 31 \\ \hline \end{array}$$
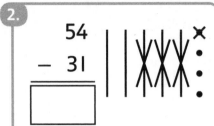

3.
$$\begin{array}{r} 35 \\ -\ 13 \\ \hline \end{array}$$

4.
$$\begin{array}{r} 56 \\ -\ 24 \\ \hline \end{array}$$

5.
$$\begin{array}{r} 66 \\ -\ 4 \\ \hline \end{array}$$

6.
$$\begin{array}{r} 49 \\ -\ 30 \\ \hline \end{array}$$

7.
$$\begin{array}{r} 87 \\ -\ 46 \\ \hline \end{array}$$

8.
$$\begin{array}{r} 95 \\ -\ 61 \\ \hline \end{array}$$

◯ Subtract.

9.

	8	5
−	4	2
	4	3

   8 tens 5 ones
− 4 tens 2 ones
  4 tens 3 ones

10.

	6	7
−	2	5

   6 tens 7 ones
− 2 tens 5 ones
___ tens ___ones

11.

	9	7
−	2	1

   9 tens 7 ones
− 2 tens 1 one
___ tens ___ ones

12.

	7	3
−	4	2

   7 tens 3 ones
− 4 tens 2 ones
___ tens ___ one

☐ Subtract.
☐ Check your answer by adding.

13.

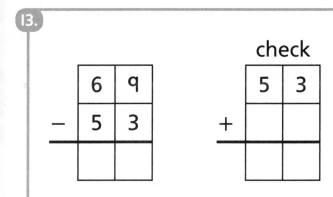

	6	9
−	5	3

check

	5	3
+		

14.

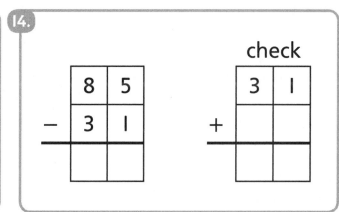

	8	5
−	3	1

check

	3	1
+		

15.

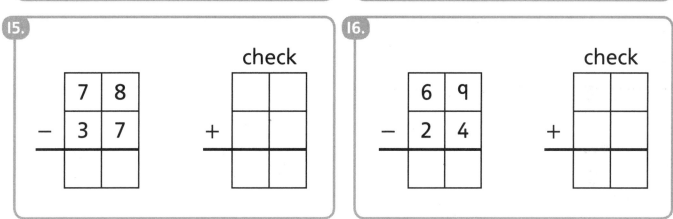

	7	8
−	3	7

check

+		

16.

	6	9
−	2	4

check

+		

☐ Subtract the ones.
☐ Keep the tens the same.

17.

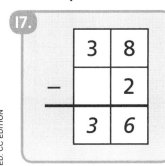

	3	8
−		2
	3	6

18.

	7	9
−		5

19.

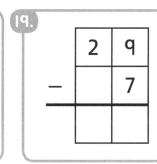

	2	9
−		7

20.

	5	5
−		2

21.

	8	6
−		3

22.

	6	7
−		1

23.

	1	6
−		4

24.

	3	8
−		8

⬭ Subtract the ones. Then subtract the tens.

25.
4	5
− 2	1
2	4

26.
3	4
− 1	2

27.
5	8
− 2	3

28.
4	4
− 3	0

29.
6	9
− 1	8

30.
7	8
− 3	5

31.
8	9
− 1	9

32.
5	6
− 4	3

33.
6	1
− 5	0

34.
7	9
− 4	4

35.
8	7
− 4	3

36.
3	3
− 2	1

37.
9	7
− 5	4

38.
8	5
− 3	2

39.
9	4
− 2	1

40.
6	5
− 4	3

41. BONUS
8	7
−	
6	1

42. BONUS
9	5
−	
4	4

43. BONUS
6	9
−	
2	3

44. BONUS
9	8
−	
3	2

Number and Operations in Base Ten 2-21

NBT2-22 Regrouping for Subtraction

☐ What number does the picture show?

1.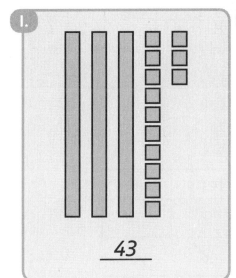

_43_

2.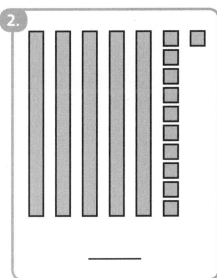

_____

3.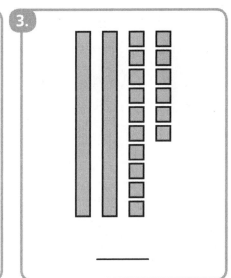

_____

☐ Take apart a ten.

4.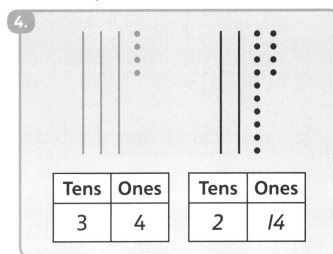

Tens	Ones
3	4

Tens	Ones
2	14

5.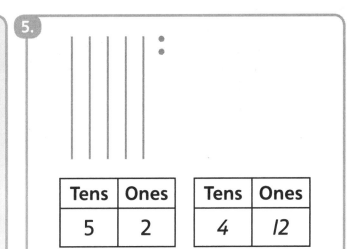

Tens	Ones
5	2

Tens	Ones
4	12

6.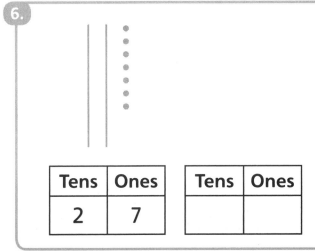

Tens	Ones
2	7

Tens	Ones

7.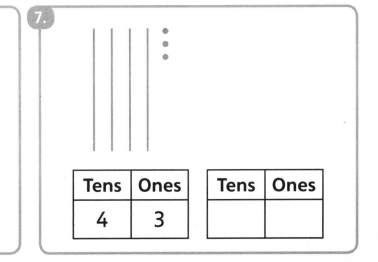

Tens	Ones
4	3

Tens	Ones

☐ Take apart a ten.

☐ Show the change in the tens and ones chart.

8.

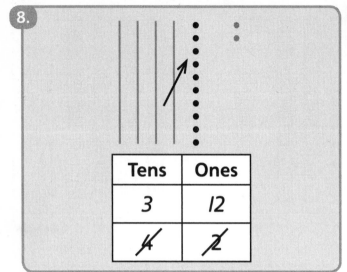

Tens	Ones
3	12
~~4~~	~~2~~

9.

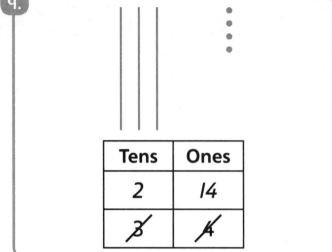

Tens	Ones
2	14
~~3~~	~~4~~

10.

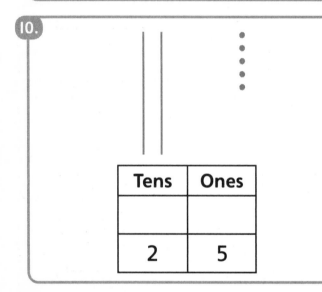

Tens	Ones
2	5

11.

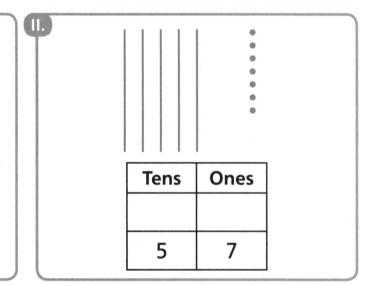

Tens	Ones
5	7

12.

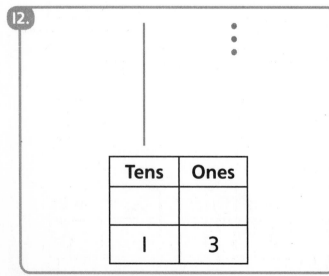

Tens	Ones
1	3

13.

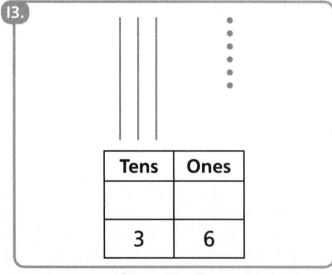

Tens	Ones
3	6

◯ Take away 1 ten from the tens. Add 10 ones to the ones.
◯ Show the change in the tens and ones chart.

14.

$50 = \underline{\quad 5 \quad}$ tens $+ \underline{\quad 0 \quad}$ ones

$ = \underline{\quad 4 \quad}$ tens $+ \underline{\quad 10 \quad}$ ones

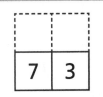

15.

$73 = \underline{\quad 7 \quad}$ tens $+ \underline{\quad 3 \quad}$ ones

$ = \underline{\qquad}$ tens $+ \underline{\qquad}$ ones

16.

$85 = \underline{\qquad}$ tens $+ \underline{\qquad}$ ones

$ = \underline{\qquad}$ tens $+ \underline{\qquad}$ ones

17.

18.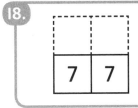

19.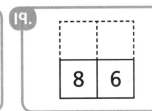

20.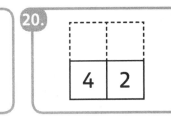

21.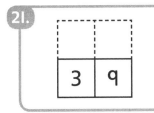

22.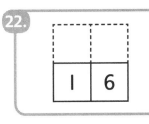

23.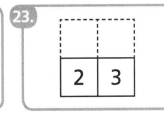

24.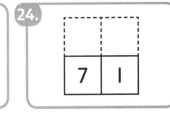

25.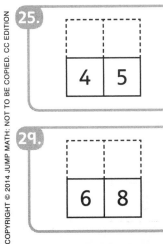

26.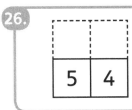

27.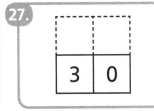

28.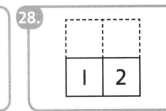

29.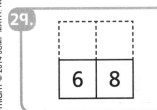

30.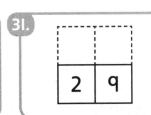

31.

32.

NBT2-23 Using Place Value to Subtract (Regrouping)

☐ Subtract.

1.

	6	15
	7̷	5̷
−	5	7
	1	8

2.

	8	3
−	5	6

3.

	5	4
−	3	9

4.

	4	6
−	2	7

5.

	9	2
−	8	7

6.

	8	1
−	5	5

7.

	3	3
−	2	9

8.

	4	0
−	3	6

9.

	8	1
−	7	2

10.

	3	6
−	2	9

11.

	4	7
−	3	8

12.

	6	0
−	5	7

☐ Regroup only if you need to.
☐ Subtract.

13.

	4	8
−	2	5

14.

	4	7
−	1	9

15.

	4	9
−	1	7

16.

	5	3
−	4	8

17.

	5	8
−	4	3

18.

	6	7
−	3	3

19.

	5	8
−	2	6

20.

	7	0
−	3	7

21.

	8	1
−	6	1

22.

	9	8
−	2	7

23.

	4	5
−	3	6

24.

	9	0
−	4	8

⬭ Regroup only if you need to.
⬭ Subtract.

25.
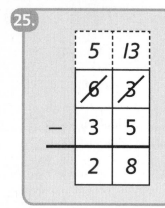

	5	13
	̶6̶	̶3̶
−	3	5
	2	8

26.
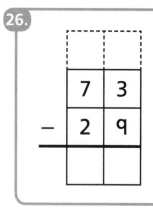

	7	3
−	2	9

27.
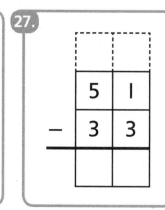

	5	1
−	3	3

28.

	8	4
−	6	7

29.
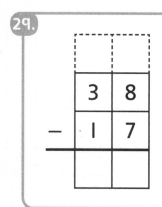

	3	8
−	1	7

30.
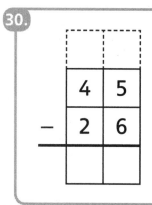

	4	5
−	2	6

31.
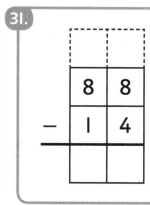

	8	8
−	1	4

32.

	2	6
−	1	9

33.
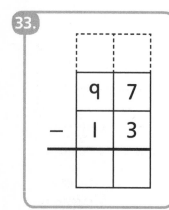

	9	7
−	1	3

34.
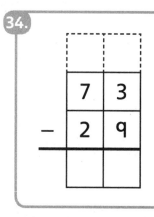

	7	3
−	2	9

35.
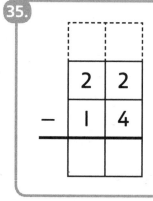

	2	2
−	1	4

36.

	8	6
−	6	1

37.

Rita did this subtraction.

What mistake did she make?

	1	5
	6	̶5̶
−	3	8
	3	7

Number and Operations in Base Ten 2-23

NBT2-24 Subtraction Word Problems

☐ Subtract.

1.

There were 27 raisins in a box.

Abdul ate 13 raisins.

How many raisins are left?

```
    2   7
-   1   3
    1   4
```

2.

93 bees were in a hive.

44 bees flew away.

How many bees stayed in the hive?

3.

86 ants marched in a line.

Some ants wandered off. 52 kept marching.

How many ants wandered off?

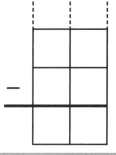

4.

Blanca had 75 pennies.

She put some in a jar. Now she has 39 pennies.

How many pennies did she put in the jar?

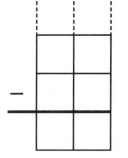

5. BONUS

Amy ate 25 grapes. There were 42 grapes before.

How many grapes are left?

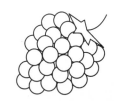

☐ Subtract.

6.

Carl has 25 apples.

12 apples are red. The rest are green.

How many apples are green?

	2	5
−	1	2

7.

57 guppies are in a fish tank.

29 are black. The rest are blue.

How many are blue?

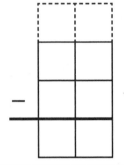

−		

8.

Jen has 21 baseball cards.

Mike has 9 baseball cards.

How many more baseball cards does Jen have?

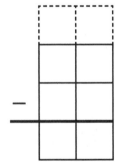

−		

9.

Anwar has 23 stickers.

Vicki has 17 stickers.

How many more stickers does Anwar have?

10. BONUS

There are 24 blue pencils and 49 red pencils in a box.

How many more red pencils are there?

MD2-I Length, Width, and Height

☐ Color the **longer** pencil.

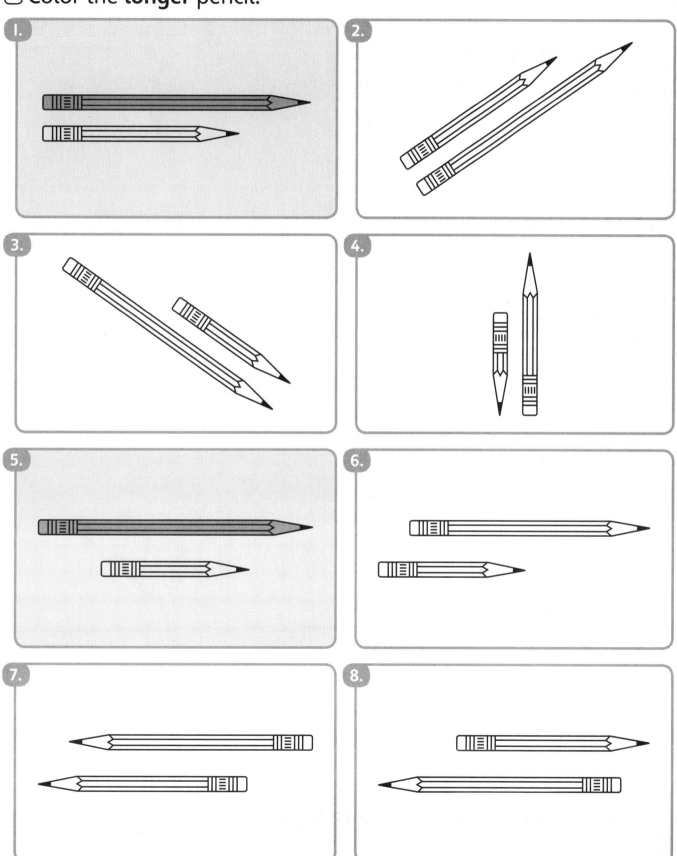

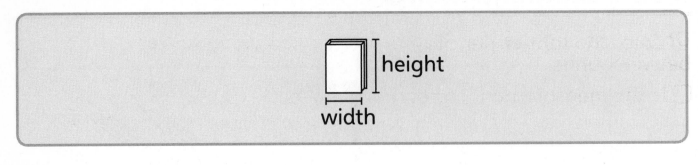

☐ Color the line showing **width** blue.
☐ Color the line showing **height** red.

9.

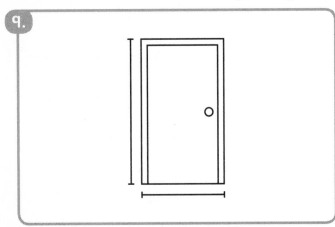

10.

11.

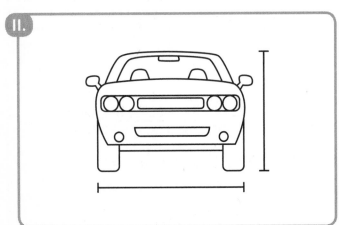

12.

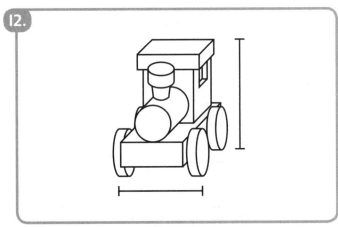

13. BONUS

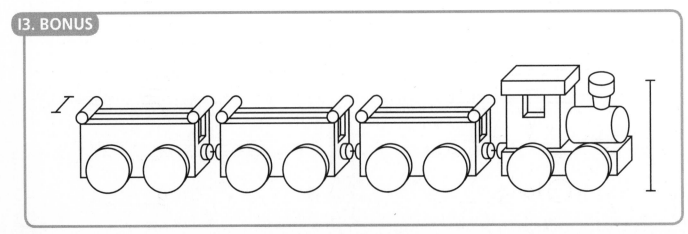

MD2-2 Measuring Length

Units must be the same length. There must be no spaces between units.

☐ Is the measurement correct? Write ✓ or ✗.

1.

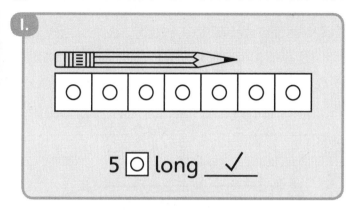

5 ◻ long ___✓___

2.

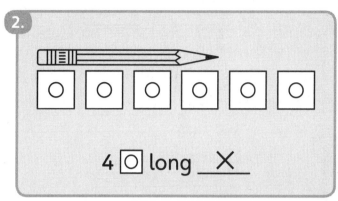

4 ◻ long ___✗___

3.

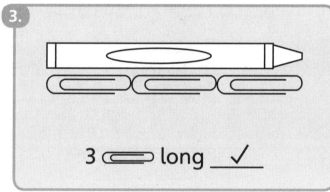

3 ⬭ long ___✓___

4.

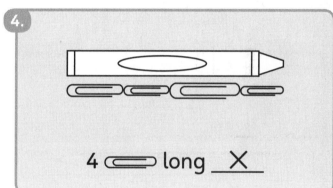

4 ⬭ long ___✗___

5.

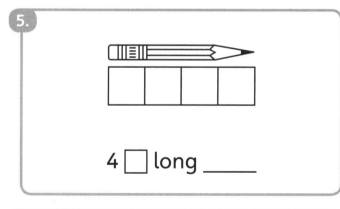

4 ☐ long _____

6.

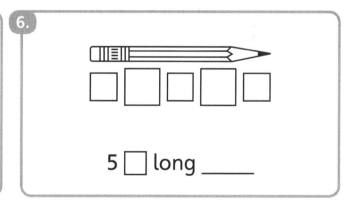

5 ☐ long _____

7.

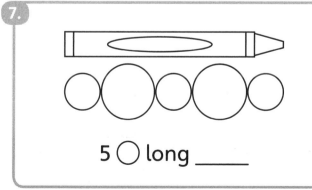

5 ◯ long _____

8.

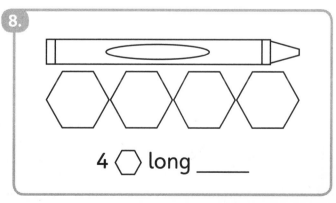

4 ⬡ long _____

☐ Is the measurement correct? Write ✓ or ✗.

9.

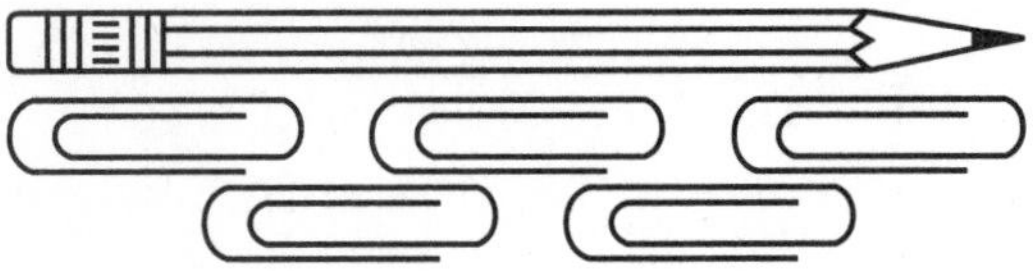

5 ⬭ long _____

10.

3 ⬭ long _____

11.

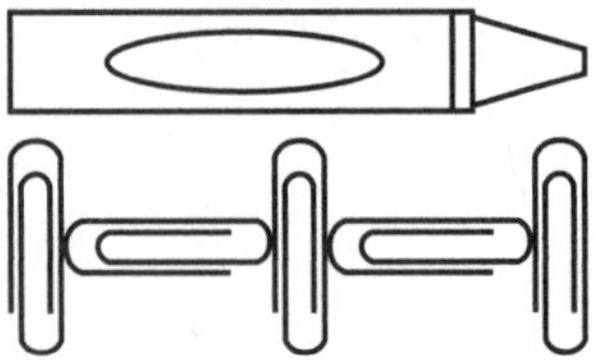

5 ⬭ long _____

12.

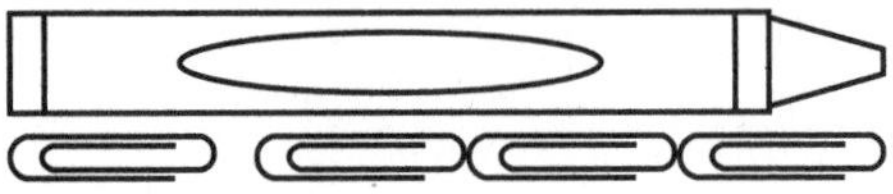

4 ⬭ long _____

13.

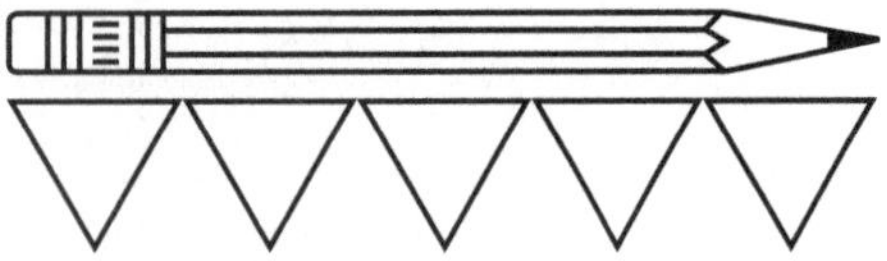

5 ▽ long _____

14.

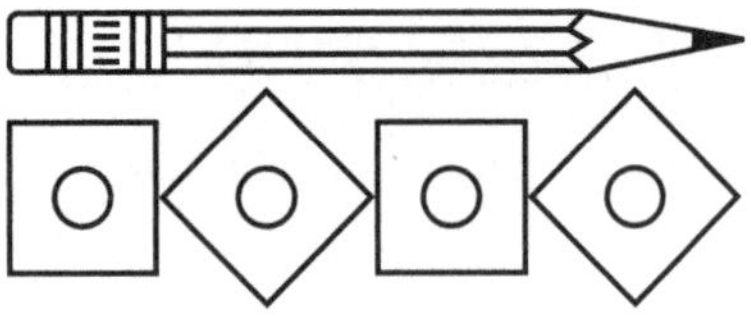

4 ▣ long _____

15.

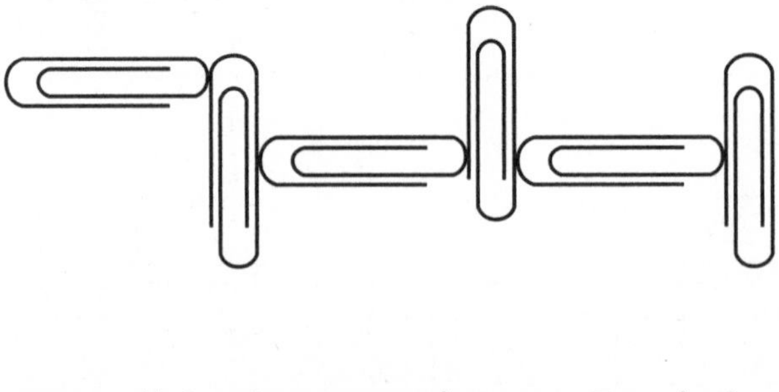

Kim says the pencil is 6 ⬭ long. Explain her mistake.

_____

Measurement and Data 2-2

MD2-3 Measuring in Centimeters

A  is I **centimeter** long.

☐ Write how many centimeters long.

1.

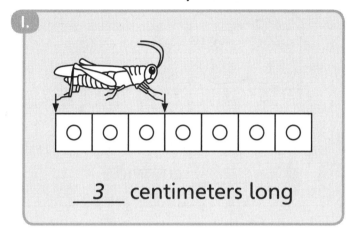

___3___ centimeters long

2.

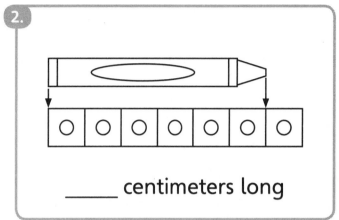

_____ centimeters long

3.

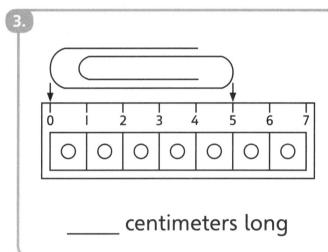

_____ centimeters long

4.

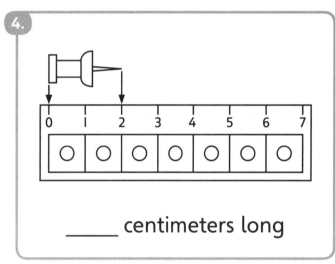

_____ centimeters long

5.

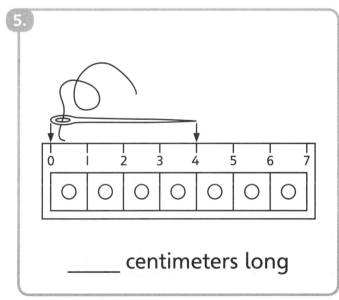

_____ centimeters long

6.

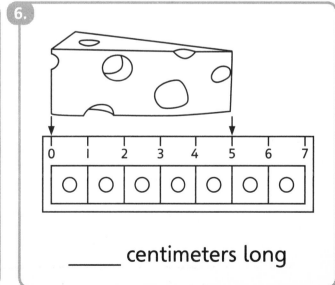

_____ centimeters long

We can write **cm** for **centi**meter.

☐ Measure how many cm.

7.

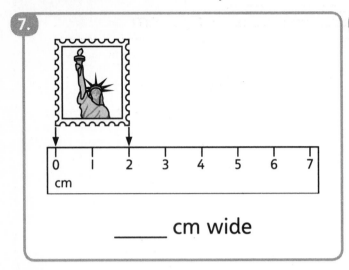

_____ cm wide

8.

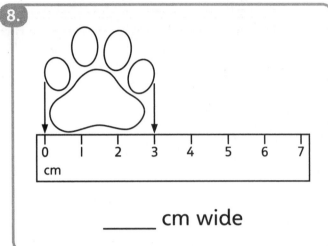

_____ cm wide

9.

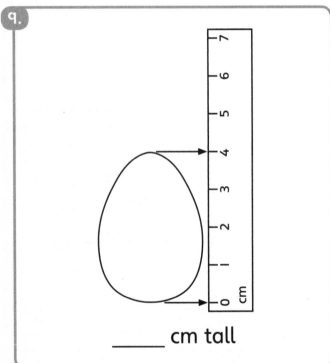

_____ cm tall

10.

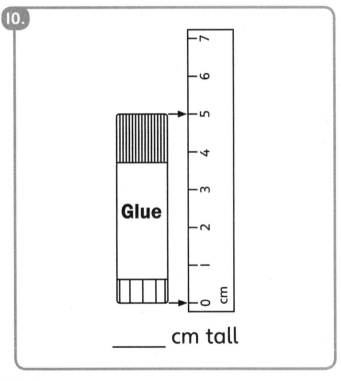

_____ cm tall

11.

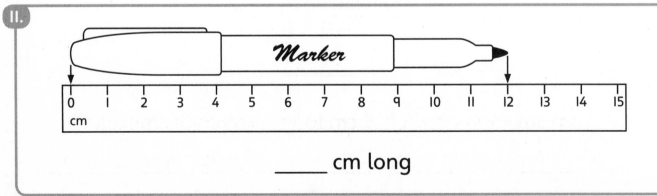

_____ cm long

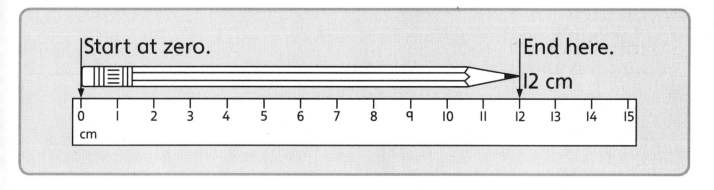

Start at zero.   End here.

12 cm

☐ Measure the object.

12.

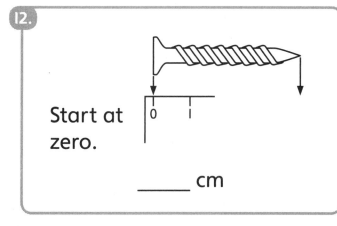

Start at zero.

_____ cm

13.

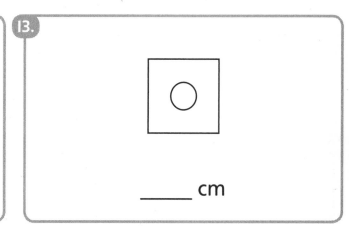

_____ cm

14.

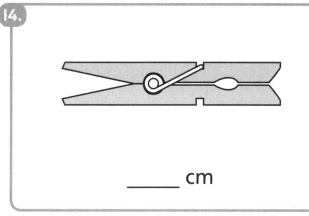

_____ cm

15.

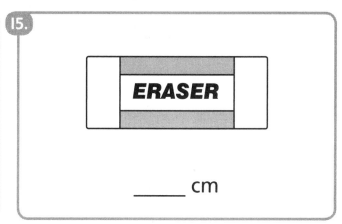

ERASER

_____ cm

16.

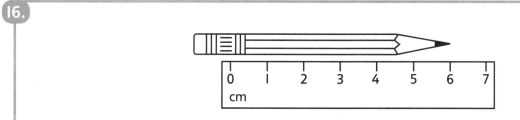

Sam says the pencil is 6 cm long. Explain his mistake.

_____

MD2-4 Length and Subtraction

How far apart are the arrows?
◯ Count the jumps, starting at zero.

1.
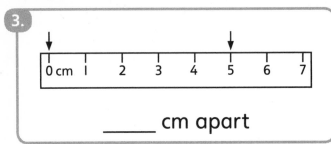
____4____ cm apart

2.
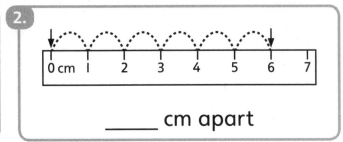
_____ cm apart

3.
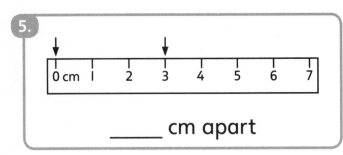
_____ cm apart

4.
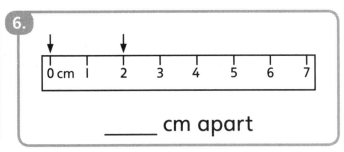
_____ cm apart

5.
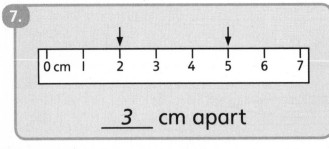
_____ cm apart

6.
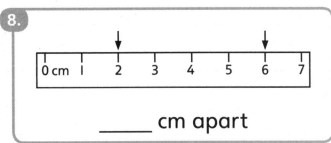
_____ cm apart

How far apart are the arrows?
◯ Count the jumps.

7.
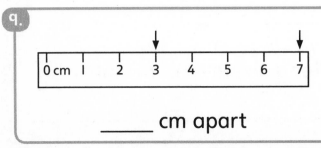
____3____ cm apart

8.
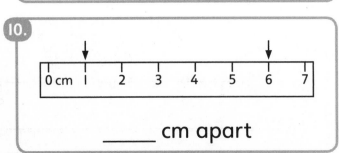
_____ cm apart

9.
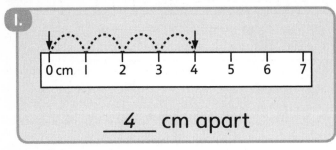
_____ cm apart

10.
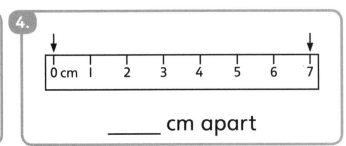
_____ cm apart

◯ Measure the length of the line or object.

11.

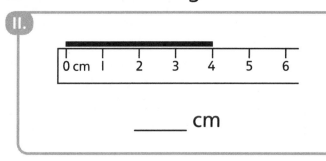

_____ cm

12.

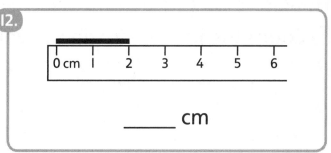

_____ cm

13.

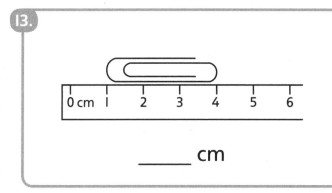

_____ cm

14.

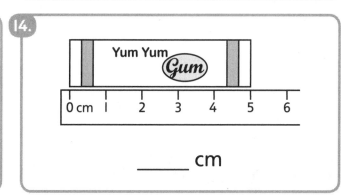

Yum Yum *Gum*

_____ cm

15.

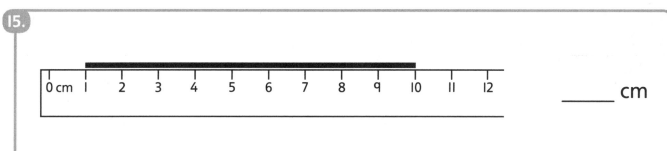

_____ cm

16.

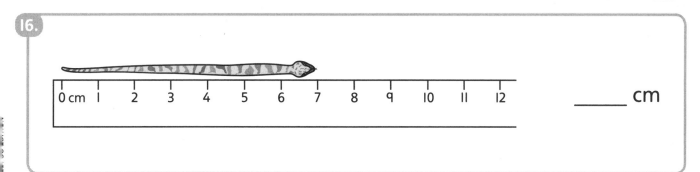

_____ cm

17.

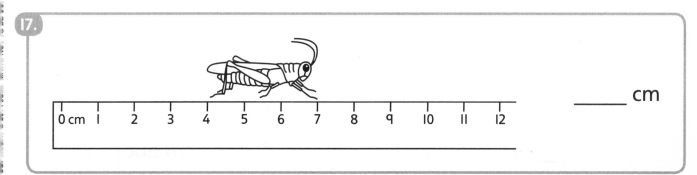

_____ cm

Bo counts jumps to find the length.

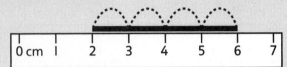

The line is __4__ cm long.

Jen subtracts to find the length.

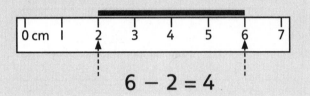

$6 - 2 = 4$

The line is __4__ cm long.

⬭ Subtract to find the length.

18.

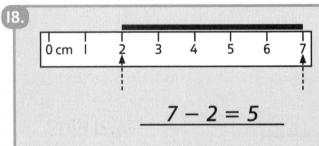

$7 - 2 = 5$

The line is __5__ cm long.

19.

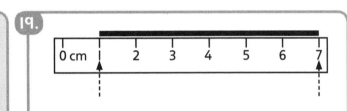

_____

The line is _____ cm long.

20.

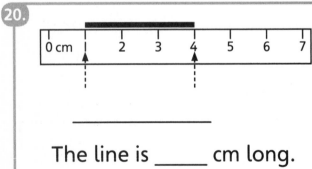

_____

The line is _____ cm long.

21.

_____

The line is _____ cm long.

22.

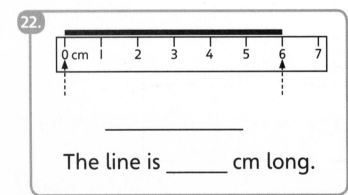

_____

The line is _____ cm long.

23.

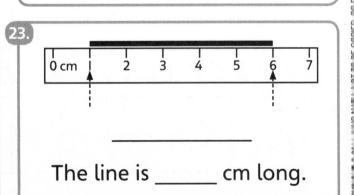

_____

The line is _____ cm long.

MD2-5 Measuring to the Closest Centimeter

☐ Which measurement is the length closest to?

1.

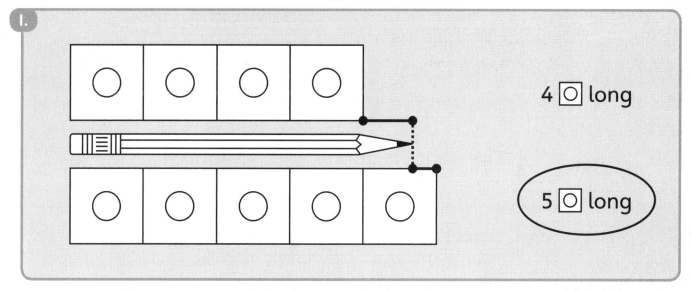

4 ☐ long

5 ☐ long

2.

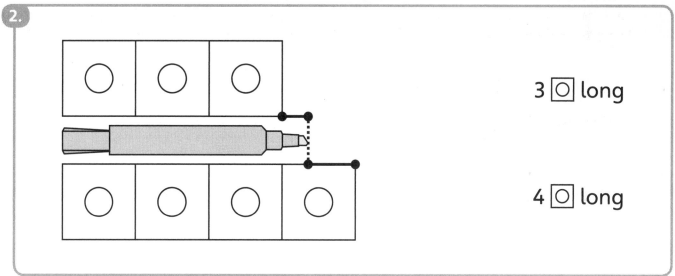

3 ☐ long

4 ☐ long

3.

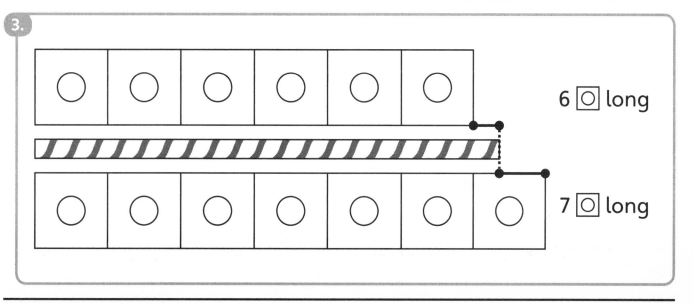

6 ☐ long

7 ☐ long

☐ How long is the pencil?

4.

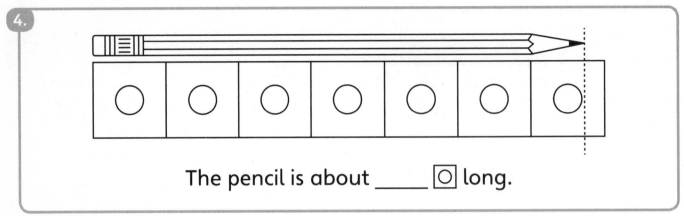

The pencil is about _____ ☐ long.

5.

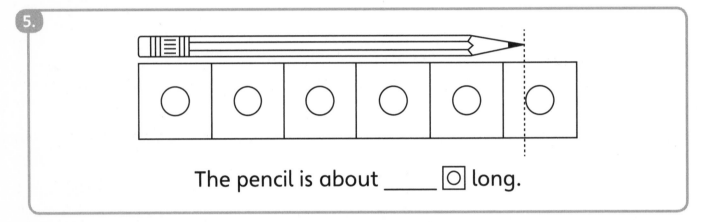

The pencil is about _____ ☐ long.

6.

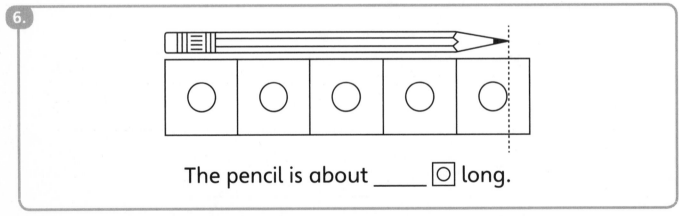

The pencil is about _____ ☐ long.

7.

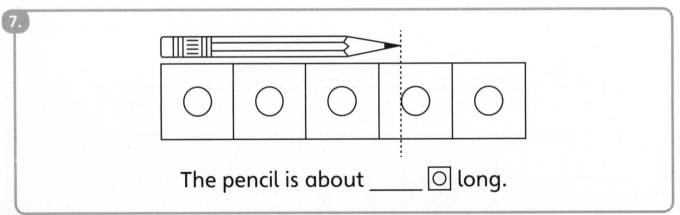

The pencil is about _____ ☐ long.

☐ How long is the pencil?

8.

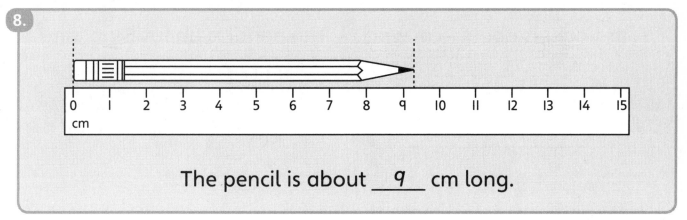

The pencil is about __9__ cm long.

9.

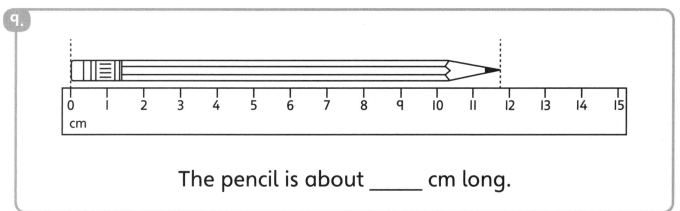

The pencil is about _____ cm long.

10.

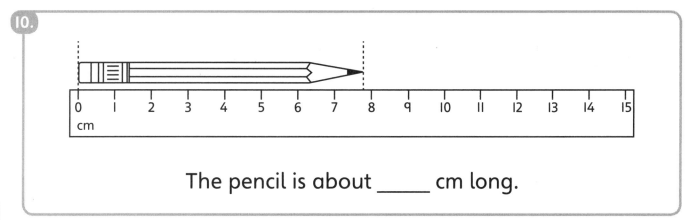

The pencil is about _____ cm long.

11.

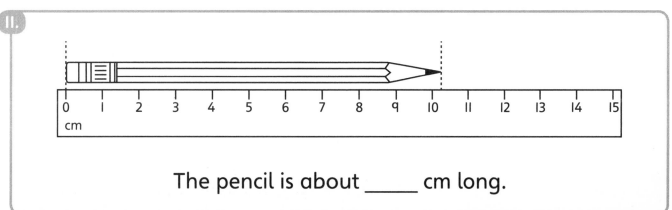

The pencil is about _____ cm long.

MD2-6 Estimating in Centimeters

Your finger is about 1 cm wide.   The pencil is about 6 cm long.

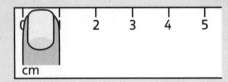 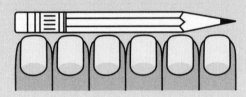

☐ Use your finger to estimate the length.
☐ Measure the length.

Object	Estimate	Measurement
(push pin)	about _____ cm	_____ cm
ERASER	about _____ cm	_____ cm
(paper clip)	about _____ cm	_____ cm
(pencil)	about _____ cm	_____ cm
(crayon)	about _____ cm	_____ cm

MD2-7 Estimating in Meters

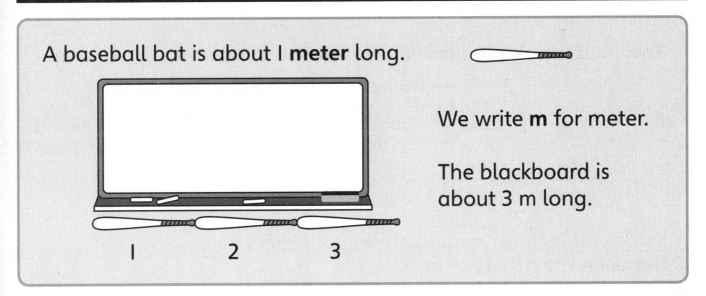

A baseball bat is about 1 **meter** long.

We write **m** for meter.

The blackboard is about 3 m long.

1   2   3

☐ Estimate the **width** to the nearest meter.
☐ Measure the width.

1.

Object	Estimate	Measurement
A door	about _____ m	_____ m
Your classroom	about _____ m	_____ m
A window	about _____ m	_____ m
The hallway	about _____ m	_____ m
The gym	about _____ m	_____ m

MD2-8 Making a Ruler

The marks on the ruler are the same distance apart.

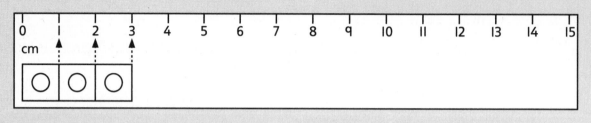

☐ Which rulers have marks with equal spaces between them? Use ✓ or ✗.

1.

Make marks with equal spaces between them.

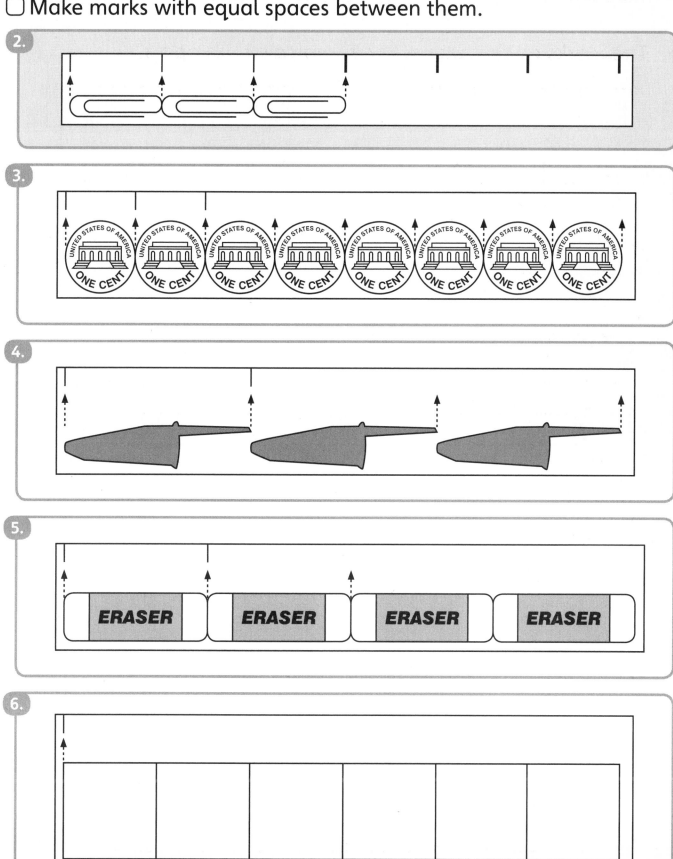

☐ Fill in the missing numbers on the ruler.

7.

0 1 2 3 4

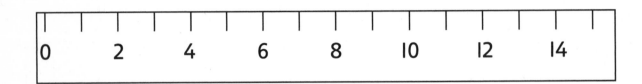

0   2   4   6   8   10   12   14

0   1   2

0   2   4   6

1   3   5   7   9   11   13   15

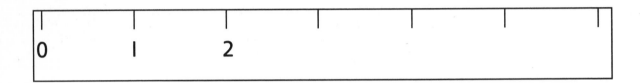

1   3   5

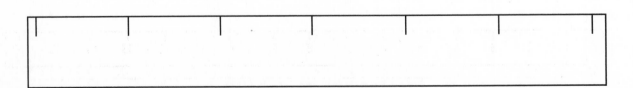

0   3

Measurement and Data 2-8

MD2-9 Drawing Lengths

Ed wants to draw a line 5 cm long.
He draws a dot at the zero mark on the ruler.

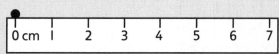

He counts on 5 cm and draws the second dot.

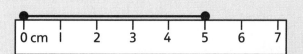

Ed connects the dots.

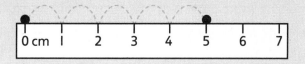

☐ Draw a second dot. Connect the dots.

1.
4 cm apart

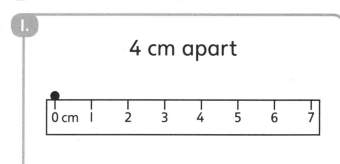

2.
3 cm apart

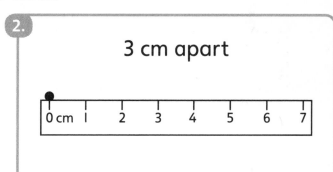

3.
6 cm apart

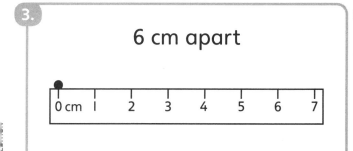

4.
2 cm apart

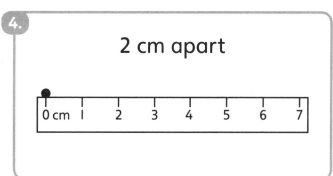

5.
10 cm apart

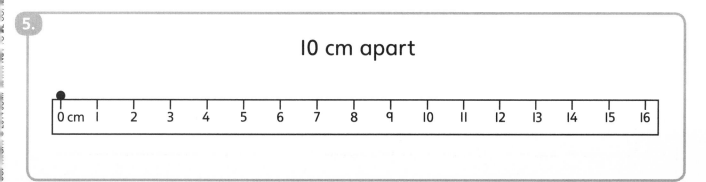

☐ Draw a second dot. Connect the dots.

6.
5 cm
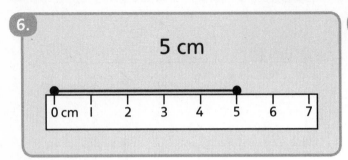

7.
4 cm
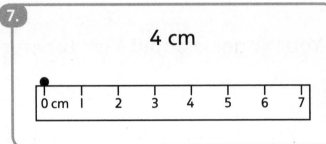

8.
3 cm
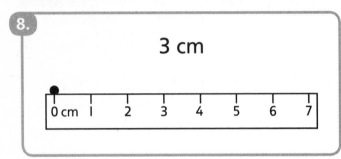

9.
6 cm
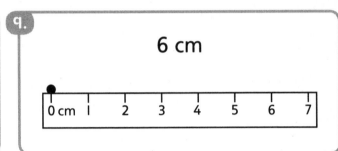

☐ Draw a line to show the length.

10.
3 cm
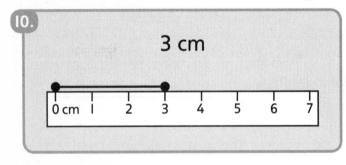

11.
7 cm
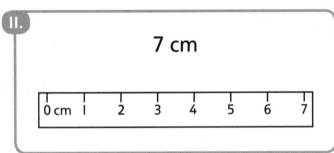

12.
4 cm
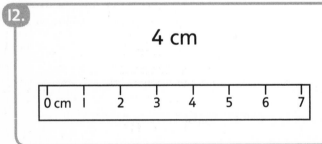

13.
2 cm
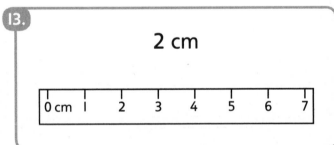

☐ **BONUS:** Start at the I cm mark.

14.
3 cm
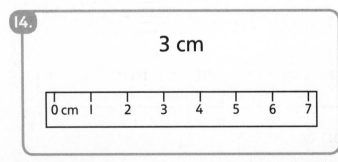

15.
6 cm
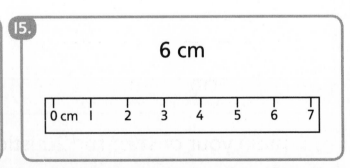

MD2-I0 Choosing Units

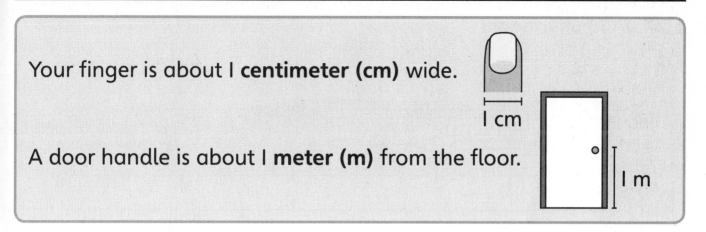

Your finger is about I **centimeter (cm)** wide.

I cm

A door handle is about I **meter (m)** from the floor.

I m

☐ Which unit would you use to measure length? Circle **cm** or **m**.

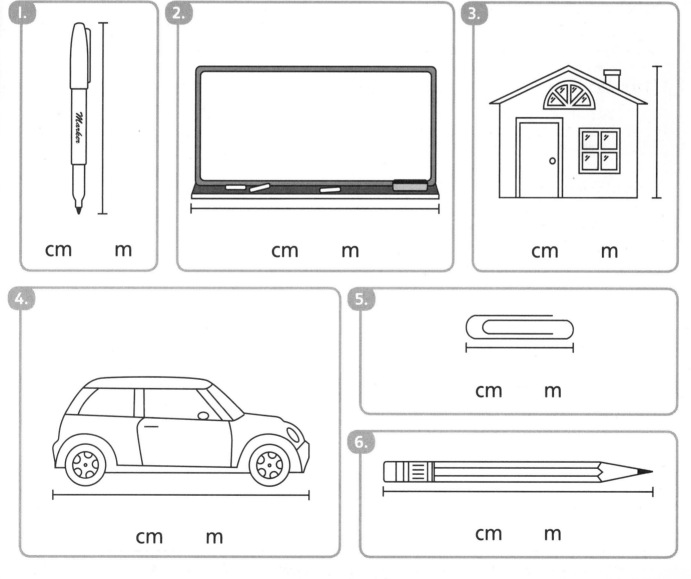

1.

cm   m

2.

cm   m

3.

cm   m

4.

cm   m

5.

cm   m

6.

cm   m

☐12. Explain your answer for Question 6.

MD2-II Comparing Lengths

☐ How much longer?

1.

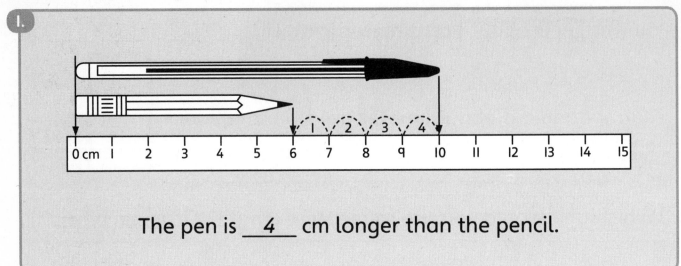

The pen is __4__ cm longer than the pencil.

2.

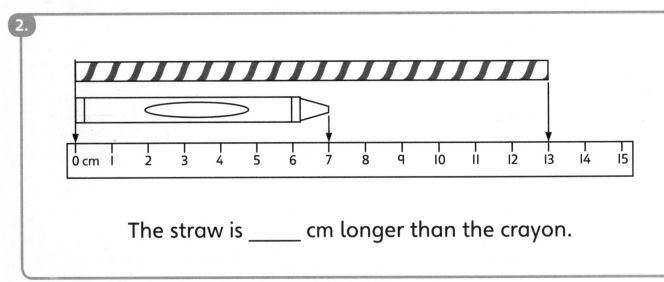

The straw is _____ cm longer than the crayon.

3.

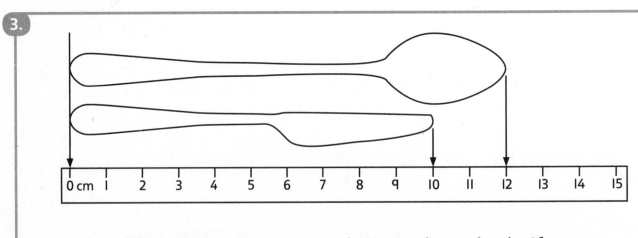

The spoon is _____ cm longer than the knife.

⬜ How much longer?

4.

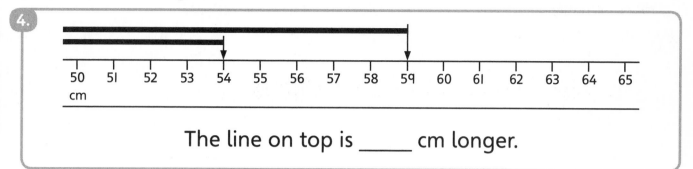

The line on top is _____ cm longer.

5.

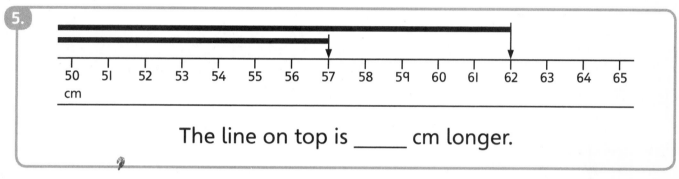

The line on top is _____ cm longer.

6.

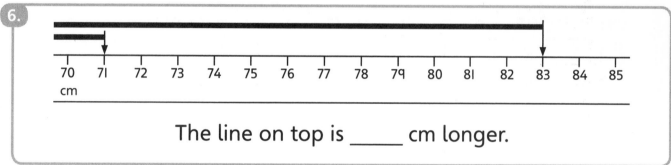

The line on top is _____ cm longer.

7.

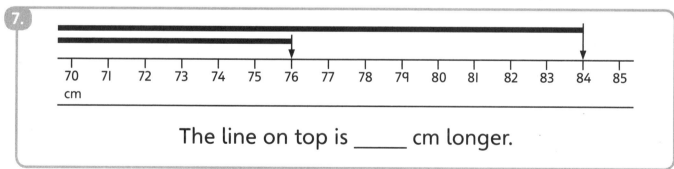

The line on top is _____ cm longer.

8.

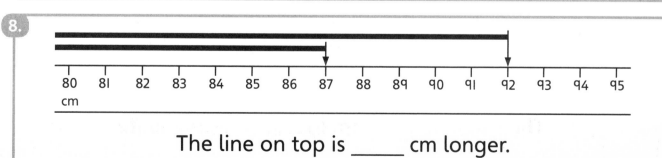

The line on top is _____ cm longer.

MD2-12 Subtraction and Length

☐ How much longer? Write a subtraction sentence.

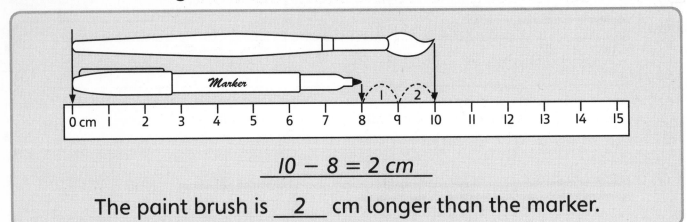

$$10 - 8 = 2 \text{ cm}$$

The paint brush is __2__ cm longer than the marker.

1.
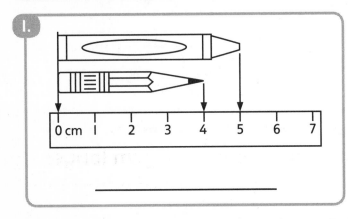

_____

2.
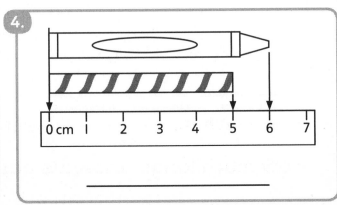

_____

3.
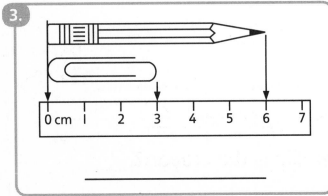

_____

4.
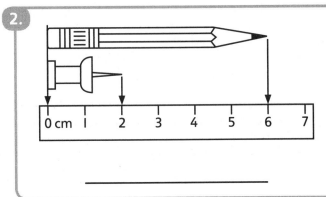

_____

5.
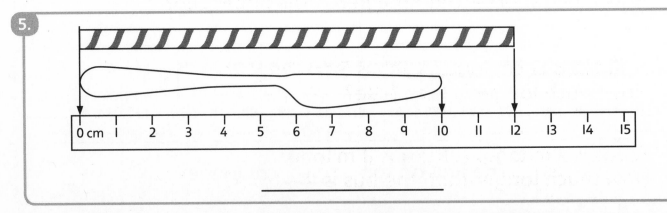

_____

Measure the animals.
How much longer is the worm? Show your work.

6.

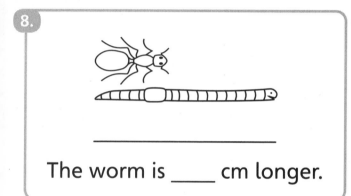

$5 - 4 = 1\ cm$

The worm is __1__ cm longer.

7.

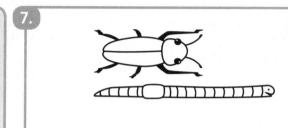

_____

The worm is ____ cm longer.

8.

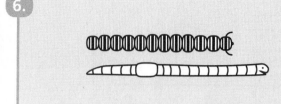

_____

The worm is ____ cm longer.

9.

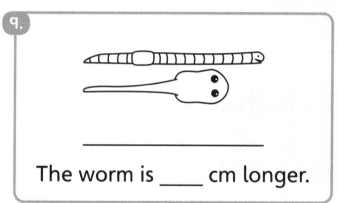

_____

The worm is ____ cm longer.

Use the measurements to answer the questions.

10.

crayon	paper clip	pen	pencil
5 cm	3 cm	8 cm	10 cm

How much longer than the pen is the pencil? _____

How much longer than the paper clip is the crayon? _____

How much shorter than the pen is the paper clip? _____

11. A turtle is 12 cm long. A fish is 5 cm long.
How much longer is the turtle?

12. A bus is 7 m long. A truck is 11 m long.
How much longer than the bus is the truck?

MD2-13 Addition and Length

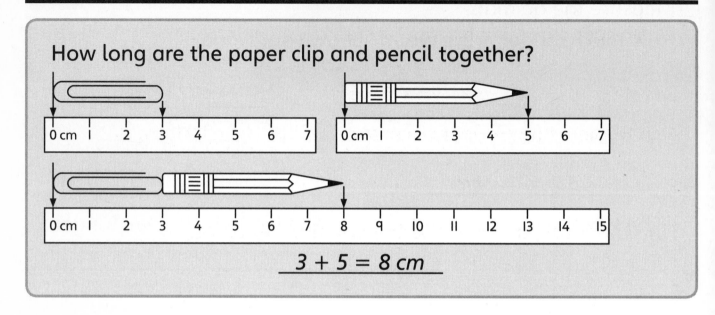

How long are the paper clip and pencil together?

$$3 + 5 = 8 \text{ cm}$$

☐ Draw an arrow where the pencil ends.
☐ Write the addition sentence.

1.

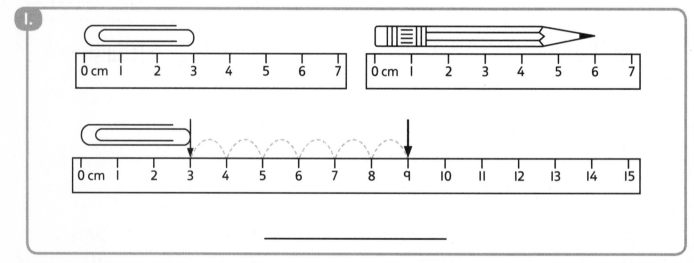

_____

2.

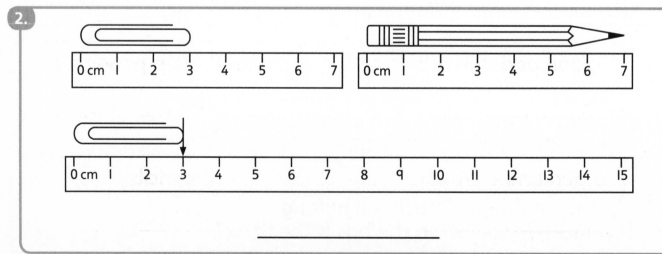

_____

The paper clip is 3 cm long. The tack is 2 cm long.

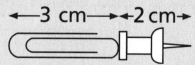

←—3 cm—→ ←2 cm→

Emma writes an addition sentence for the total length.

$3 + 2 = 5$ cm

⬜ Find the total length. Write the addition sentence.

3.

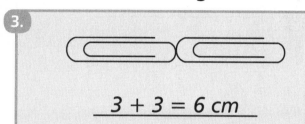

$3 + 3 = 6$ cm

4.

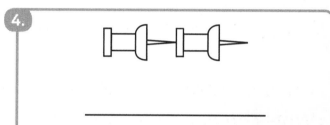

_____

5.

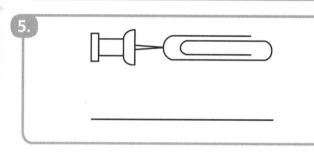

_____

6.

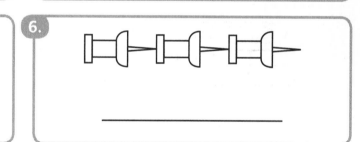

_____

⬜ Find the total length. Write the addition sentence.

crayon	paper clip	pen	pencil
5 cm	3 cm	8 cm	10 cm

7. a pen and a pencil

_____

8. a pencil and a paper clip

_____

9. a pen and a crayon

_____

10. 2 pencils

_____

MD2-14 Addition and Length (Advanced)

◯ Measure the parts of the string.
◯ Write an addition sentence to find the total length.

1.
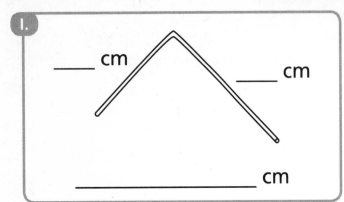
_____ cm

_____ cm

_____ cm

2.
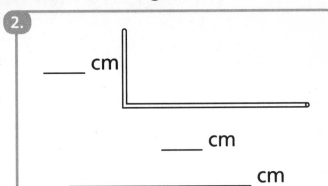
_____ cm

_____ cm

_____ cm

3.
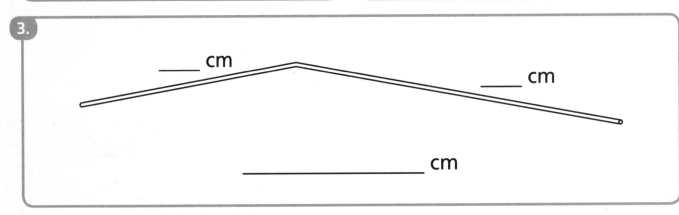
_____ cm

_____ cm

_____ cm

4.
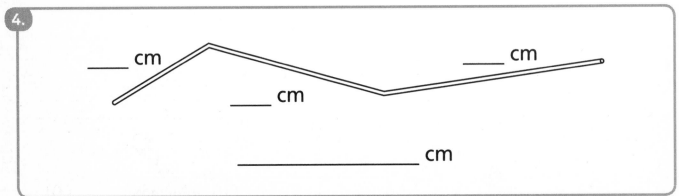
_____ cm

_____ cm

_____ cm

_____ cm

5. BONUS

A table is 2 m long.

Write an addition sentence for each length.

2 tables	3 tables	4 tables
_2 + 2 = 4 m_	_____	_____

Find the total distance the ant travels.

6.

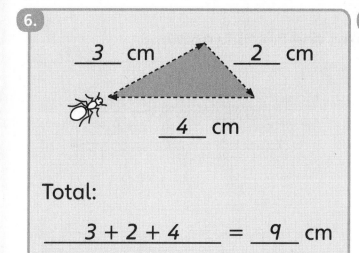

__3__ cm   __2__ cm

__4__ cm

Total:

_____3 + 2 + 4_____ = __9__ cm

7.

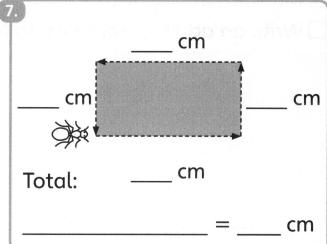

____ cm

____ cm   ____ cm

Total: ____ cm

_____ = ____ cm

8.

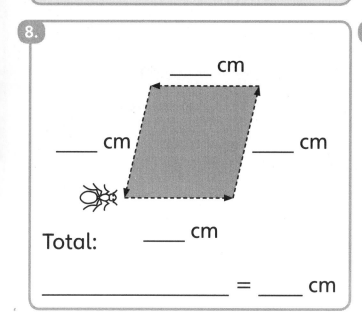

____ cm

____ cm   ____ cm

Total: ____ cm

_____ = ____ cm

9.

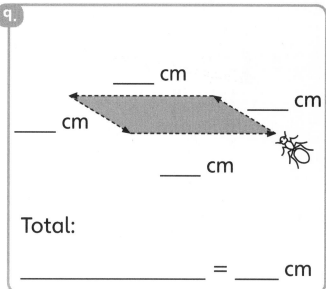

____ cm

____ cm   ____ cm

____ cm

Total:

_____ = ____ cm

10.

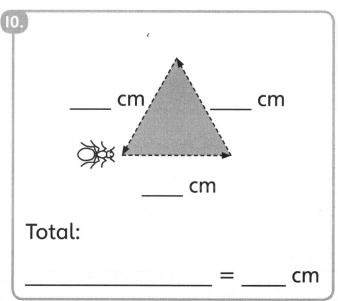

____ cm   ____ cm

____ cm

Total:

_____ = ____ cm

11.

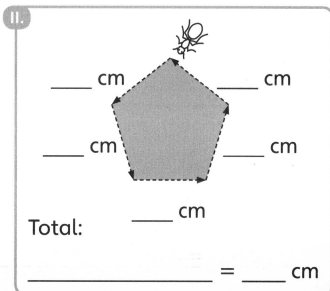

____ cm   ____ cm

____ cm   ____ cm

____ cm

Total:

_____ = ____ cm

MD2-15 Finding Lengths from Differences

The screw is 4 cm long.

☐ Write an addition sentence for the length of the nail.

1.
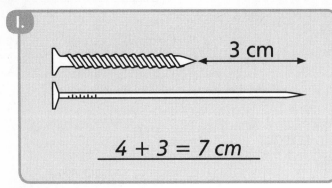
$$4 + 3 = 7 \text{ cm}$$

2.
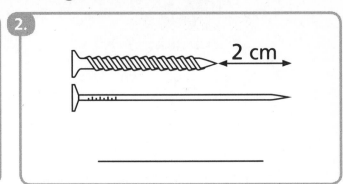
_____

3.
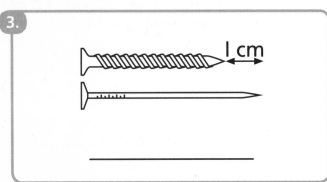
_____

4.
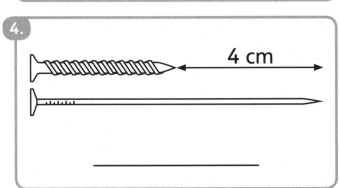
_____

☐ The screw is 3 cm long. Find the length of the nail.

☐ Check by measuring.

5.
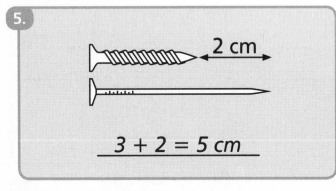
$$3 + 2 = 5 \text{ cm}$$

6.
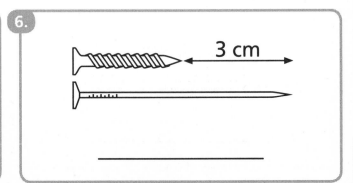
_____

7.
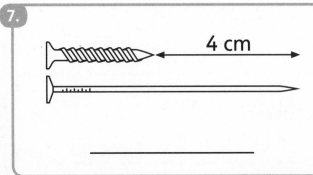
_____

8.
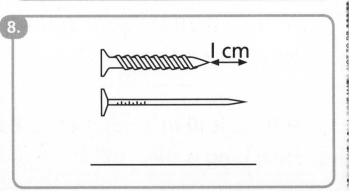
_____

☐ Draw the line.

9. 2 cm longer than the screw

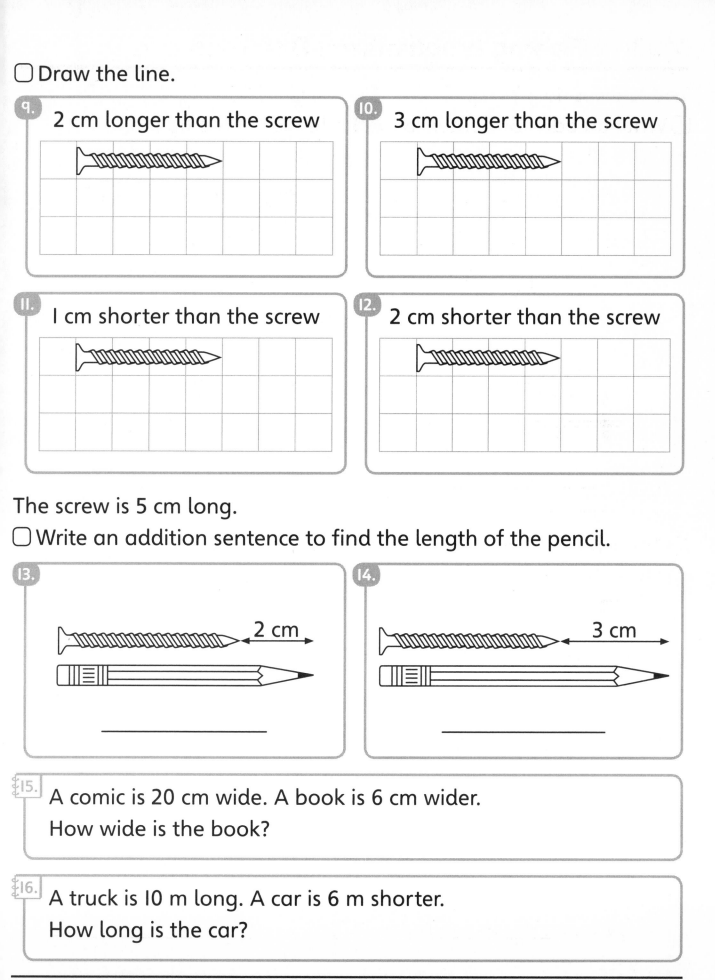

10. 3 cm longer than the screw

11. I cm shorter than the screw

12. 2 cm shorter than the screw

The screw is 5 cm long.

☐ Write an addition sentence to find the length of the pencil.

13.

2 cm

_____

14.

3 cm

_____

15. A comic is 20 cm wide. A book is 6 cm wider.
How wide is the book?

16. A truck is I0 m long. A car is 6 m shorter.
How long is the car?

MD2-16 Solving Problems

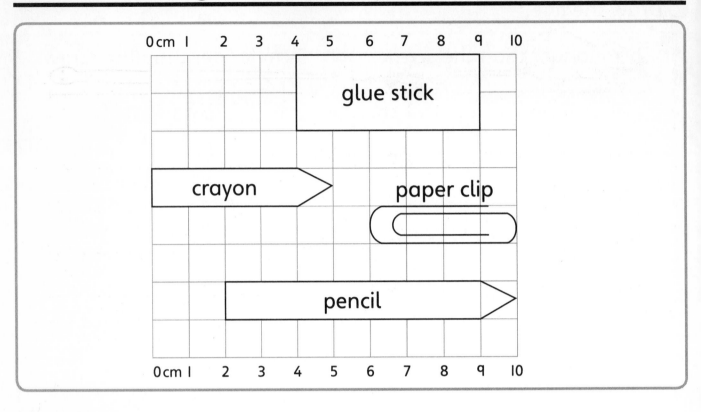

☐ How many cm long?

1.

crayon   glue stick   paper clip   pencil

_____ cm   _____ cm   _____ cm   _____ cm

☐ If you put the objects in a row, how long would they be?

2.

crayon and paper clip _____

pencil and crayon _____

glue stick and pencil _____

paper clip and glue stick _____

3. BONUS

crayon, pencil, and glue stick _____

       Measurement and Data 2-16

4.

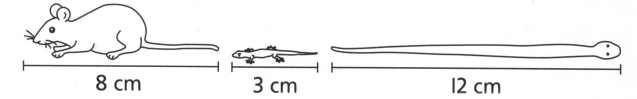

8 cm   3 cm   12 cm

How much longer than the lizard is the mouse? _____

How much longer than the mouse is the snake? _____

How much shorter than the snake is the lizard? _____

Which is the longest animal? _____

5.

Car	Length
large truck	22 m
van	6 m
small truck	10 m

How much longer than the van is the large truck? _____

How long are the van and the small truck together? _____

ee cars are parked in a row. How long is the row?

_____

2 m shorter than the van. How long is Jen's car?

_____